HLA FROM BENCHTOP TO BEDSIDE

HLA FROM BENCHTOP TO BEDSIDE

ARTHUR BRADLEY EISENBREY, III, MD, PhD

Associate Professor of Pathology,
Wayne State University School of Medicine,
Detroit, MI, United States;

Clinical Associate Professor of Pathology,
University of Toledo College of
Medicine and Life Sciences,
Toledo, OH, United States

ELSEVIER

ACADEMIC PRESS

An imprint of Elsevier

Academic Press is an imprint of Elsevier
125 London Wall, London EC2Y 5AS, United Kingdom
525 B Street, Suite 1650, San Diego, CA 92101, United States
50 Hampshire Street, 5th Floor, Cambridge, MA 02139, United States
The Boulevard, Langford Lane, Kidlington, Oxford OX5 1GB, United Kingdom

Library of Congress Cataloging-in-Publication Data
A catalog record for this book is available from the Library of Congress

British Library Cataloguing-in-Publication Data
A catalogue record for this book is available from the British Library

ISBN: 978-0-12-823976-6

For information on all Academic Press publications visit our website at https://www.elsevier.com/books-and-journals

Publisher: Stacy Masucci
Acquisitions Editor: Elizabeth Brown
Editorial Project Manager: Pat Gonzalez
Production Project Manager: Punithavathy Govindaradjane
Cover Designer: Mark Rogers

Typeset by TNQ Technologies

Contents

Preface

Understanding human histocompatibility is a challenge. Those of us who have been immersed in this study for decades marvel at the changes that have taken place in technology, the incredible diversity packed into a short stretch of the human genome, and the breadth of human health that is touched by studies in what is now known as immunogenetics. For years, those of us who teach fellows, residents, and technologists what we do in our "HLA labs" have relied on some excellent guides for introduction to the field: Marsh, Parham and Barber's excellent *The HLA FactsBook*, Academic Press, 2000, and Rodey's foundational *HLA Beyond Tears: Introduction to Human Histocompatibility*, second edition, Pel-Freez, 2000. There are many excellent review articles and textbook chapters which can provide overviews of this complex and evolving field and many will be cited in this book. Much, however, has changed since the publication of "Beyond Tears" and the "Facts Book."

This book is built on a broad foundation that began when I was a research technician for George C. Lakatos, MD, in the research labs run by Daisy McCann, PhD, at Wayne County General Hospital, Wayne, Michigan. Herta Vickery, PhD, stepped me through a crossmatch for a kidney transplant in late 1973 or early 1974. As a young research technician doing cancer immunotherapy research in mice, this looked like an exciting application of the immunology that I was learning. In retrospect, the crude agglutination of buffy coat cells from the peripheral blood of the donor with the serum of the proposed recipient was not as sophisticated as the Ouchterlony double immunodiffusion I was using for antibody detection and quantification.

The intent of this book is to provide a picture of where HLA or histocompatibility testing is four decades after my introduction to the field. Whether you are a casual reader or a student/technologist/resident/fellow trying to get a "quick understanding" of HLA, I hope that this will be an acceptable starting point for your learning. There will be a little history to put things in perspective, and then each chapter will focus on one clinically relevant subject area.

This is not a reference guide for all of the testing that may be performed in HLA testing laboratories, whether named HLA, Immunogenetics, or Transplant Immunology as my previous labs were named. Some laboratories perform infectious disease testing for organ donors, some laboratories perform non-HLA crossmatches (e.g., XM-ONE®, AbSorber AB, Stockholm, Sweden), and many laboratories that support hematopoietic stem cell transplantation perform engraftment monitoring. Most of these subjects, although related to transplantation are not HLA or Major Histocompatibility Complex (MHC) testing and are not covered in this text. Engraftment monitoring, although not HLA testing, is covered in Chapter 10 because of its importance in hematopoietic stem cell transplantation.

The references are cited for the more serious student of immunogenetics to provide more depth and to give recognition to those whose foundational work made this effort possible. I also recognize those who have taught me, mentored me, and challenged me as teachers, colleagues, supervisors, staff, resources, and friends.

This book is dedicated to the patients and donors who make transplantation a reality and to those nontransplant patients with other health issues whom we serve. Thank you for every encounter that provides a learning opportunity.

References to commercial products are not given as endorsements and are included only for informational purposes. The author has received no financial support or incentive from any vendor or supplier of clinical laboratory reagents or instrumentation.

Acknowledgments

I am eternally in debt to all of the scientists, physicians, technologists, nurses, and patients who have guided my growth as a scientist and physician studying and practicing within the broad field of immunology, particularly Daisy McCann, PhD, Herta Vickery, PhD, William S. Moore, PhD, Gerald Gleich, MD, C. Garrison Fathman, MD, Richard Walker, MD, Dorothy Levis, MT(ASCP), Sharon Skorupski, BS(CHS), all of my research collaborators, and my wife, Louise M. Eisenbrey, MS, MT(ASCP), CQM/OE(ASQ). I mark a particular note of thanks to my colleague, friend, and editor, John Gerlach, PhD, for his review, editorial comments, suggestions, and recommended changes.

In my interactions with medical students, residents, friends, colleagues, and family, I have found a tendency to "Google-it" when looking up medical information. I would like to counter that tendency with the admonition to take the extra step and use PubMed.gov (https://www.ncbi.nlm.nih.gov/pubmed/) and other government and noncommercial sources for medical and scientific information whenever possible. This book would not have been possible without PubMed and the exquisite services provided by the reference librarians and eJournal access at the Vera P. Shiffman Medical Library, Wayne State University, Detroit, MI.

CHAPTER 1

Introduction and some history

Contents

Bullet points

1. The primary human histocompatibility antigens are called human leukocyte antigens (HLAs).
2. HLA is part of the human major histocompatibility complex (MHC) located on the short arm of chromosome six.
3. Understanding of the human histocompatibility system has been evolving for 60 years and continues to evolve.

Let us get something out of the way, right up front: **HLA** is used as an acronym for human leukocyte antigen. HLA is also used as shorthand for the system of genes that code for and control the cell markers and recognition system that prevent cells and tissues from one individual from engrafting in another, even closely related, individual. A more proper name is the human major histocompatibility complex (MHC) for the very complex region on the short arm of chromosome six (6p21.31). There is much more in the MHC than genes that code for histocompatibility antigens. As we will see, later, HLA/MHC genes do much more than make it difficult to transplant organs and stem cells.

The name for the system did not start as HLA. Physician and scientist Dr. Jean-Baptiste-Gabriel-Joachim Dausset (*Nobel Prize in Physiology or Medicine,* 1980) gets credit for recognizing the existence of the new antigen system which would become known as HLA. His initial work was on antibodies in patients with leucopenia (Dausset and Nenna, 1953). He realized that there was a pattern of reactivity consistent with alloantibodies (Dausset, 1958; Carosella, 2009). Rose Payne, JJ van Rood, and others were also working on "leukoagglutinins" and had presented at conferences, but Dausset published first (Emanuel Hackel, Ph.D., personal communication).

Dausset's MAC antigen was named from the initials of the three donors whose white blood cells did not react with the leukoagglutinating antibodies he

HLA from Benchtop to Bedside
ISBN 978-0-12-823976-6
https://doi.org/10.1016/B978-0-12-823976-6.00001-9

was investigating (The antigen he named MAC eventually was renamed HLA-A2). Naming of the new antigen system was entirely by local conventions, and each of the labs investigating "leukoagglutinins" had their own names for the antigens they described. Efforts at collaboration between laboratories led to an ongoing series of workshops and conferences (http://www.ihwg.org/about/history.html and http://hla.alleles.org/nomenclature/workshops.html). At the Third Conference on Histocompatibility Testing, July 1967 (Amos, 1968), the participants took the existing and competing naming systems (FOUR, Hu, Du, LA, TO, and LC) and agreed that the first major locus recognized would be named HL-A (an apparent contraction of Hu and LA). When it later became apparent that another genetic locus was present, the system was renamed HLA, and the newly identified loci were named HLA-A and HLA-B, respectively. Antigens were named with sequential numbers, the naming convention that remains today (http://hla.alleles.org) (Robinson et al., 2015; Marsh et al., 2010).

Note: Much of the structure, function, and genetics of the human MHC was predicted from decades of work with laboratory animals, particularly mice. Knowledge of the mouse MHC (H-2) has been, and remains, foundational for understanding of the human MHC (Penn, 2002).

For students of the history of science and medicine, I recommend the very readable historical reviews of the development of the knowledge of the MHC system and HLA by Thorsby (2009) and Terasaki (2007). References for the exciting history of molecular testing and HLA will be provided in Chapters 7 and 8.

Fundamentals: innate and adaptive immunity

Some basic knowledge of the human immune system is needed to understand HLA. There are two broad groups of immune cells in all vertebrates, including *Homo sapiens*: innate immunity and adaptive immunity. The cells of the *innate* immunity system are presumed to have evolved first because the functions are shared across phyla. Generally speaking, these cells function by phagocytosis and internal lysis of pathogens or by direct cell to cell killing through the release of cytotoxins. The cells of *adaptive* immunity are shared by all vertebrates and include the two broad classes of lymphocytes (T cells and B cells), specialized proteins called immunoglobulins (membrane-bound and secreted products of B cells and their derivative plasma cells), T cell receptors (TCR), and antigen presentation by two classes of specialized membrane-bound proteins of the MHC

(Parham, 2005). The advanced and flexible functions of the adaptive immune system often rely on interactions with the cells of the innate immune system, particularly circulating mononuclear phagocytic cells and more sessile dendritic phagocytic cells.

Histocompatibility and immunity involve far more than the genes present in the MHC. Although I (semijokingly) tell students, residents, and fellows that, "The short arm of chromosome six is the most important part of the human genome," the MHC does not and cannot function without the rest of the immune system. Some critical systems that will be encountered in this book (and your other reading) are the TCR alpha and delta genes (chromosome 14q), TCR gamma (chromosome 7p), TCR beta (7q), immunoglobulin heavy chain genes (IGH; 14q), and the immunoglobulin light chains (kappa: 2p; lambda: 22q). There are cosignals, hormones, receptors, binding proteins, ligands, stabilizing protein structures, signaling peptides, and enzymes which have critical roles and will be discussed in the context of MHC-related activities.

The cellular components of human immunity are well described in hematology textbooks, textbooks of general immunology, and in online resources. The critical cells for the functions of the human MHC are lymphocytes (adaptive immunity) and the circulating and sessile mononuclear macrophages (innate immunity) which present antigen ("antigen presenting cells" or APCs) to the lymphocytes (antigen presentation will be described later). Lymphocytes are divided in two broad classes, T cells (T lymphocytes) and B cells (B lymphocytes). T cells were named for thymus-derived lymphocytes because immature or "naive" T cells were initially shown to mature in the thymus (a lymphoid tissue located in the anterior mediastinum) (Waldmann, 1979). B cells were named for the bursa of Fabricius, a gut-associated lymphoid tissue of birds, which was found to be the primary source of lymphoid cells that produce antibody (Glick, 1991). Other cell types and subtypes will be introduced as needed to describe functions related to the MHC and HLA in particular.

The following chapters in this book present broad concepts: genetics and structure, antibody detection and identification, HLA "typing" solid organ transplantation, stem cell transplantation, transfusion medicine support, and disease association/pharmacogenetics. Each topic will be presented with historical background so that the current methodologies can be seen in perspective of the changes in technology for those who have never had the experience of using the older techniques.

References

Alleles. http://hla.alleles.org/nomenclature/workshops.html.
Amos, D.B., 1968. Human histocompatibility locus HL-A. Science 159, 659–660.
Carosella, E.D., 2009. From MAC to HLA. Professor Jean Dausset, the pioneer. Hum. Immunol. 70, 661–662.
Dausset, J., 1958. Iso-leuco-anticorps. Acta Haematol. 20, 156–166.
Dausset, J., Nenna, A., 1953. Présence d'une leucoagglutinine dans trois serums de maladies leucopéniques. Sang 24, 410–417.
Glick, B., 1991. Historical perspective; the bursa of Fabricius and its influence on B-cell development, past and present. Vet. Immunol. Immunopathol. 30, 3–12.
Hla. http://hla.alleles.org.
Hwg. http://www.ihwg.org/about/history.html.
Marsh, S.G.E., Albert, E.D., Bodmer, W.F., Bontrop, R.E., Dupont, B., Erlich, H.A., Fernández-Vina, M., Geraghty, D.E., Holdsworth, R., Hurley, C.K., Lau, M., Lee, K.W., Mach, B., Mayr, W.R., Maiers, M., Müller, C.R., Parham, P., Petersdorf, E.W., Sasazuki, T., Strominger, J.L., Svejgaard, A., Terasaki, P.I., Tiercy, J.M., Trowsdale, J., 2010. Nomenclature for factors of the HLA system. Tissue Antigens 75, 291–455.
Parham, P., 2005. MHC class I molecules and KIRs in human history, health and survival. Nat. Rev. Immunol. 5, 201–214.
Penn, D.J., 2002. Major histocompatibility complex (MHC). In: eLS. John Wiley & Sons Ltd, Chichester. http://www.els.net.
Robinson, J., Halliwell, J.A., Hayhurst, J.H., Flicek, P., Parham, P., Marsh, S.G.E., 2015. The IPD and IMGT/HLA database: allele variant databases. Nucleic Acids Res. 43, D423–D431.
Terasaki, P., 2007. A brief history of HLA. Immunol. Res. 38, 139–148.
Thorsby, E., 2009. A short history of HLA. Tissue Antigens 74, 10–116.
Waldmann, H., 1979. Interactions between T and B cells: a review. J. R. Soc. Med. 72, 198–202.

CHAPTER 2

The structure of the major histocompatibility complex and major HLA components

Contents

Bullet points

1. The human major histocompatibility complex (MHC) contains the human leukocyte antigen (HLA) gene loci.
2. Class Ia loci are called "Classical" HLA loci (HLA-A, HLA-B, HLA-C).
3. Class Ia proteins are constitutively expressed on the surface of all nucleated cells.
4. Class Ib proteins have restricted tissue distribution and limited antigen-presenting repertoires.
5. Class II proteins (HLA-DR, HLA-DQ, and HLA-DP) are primarily expressed on the surface of antigen-presenting cells and are *inducible* on other nucleated cells.
6. Class I proteins present cytosolic self-antigens to CD8 T cells.
7. Class II proteins present external (cell environment) antigens to CD4 T cells.

The human major histocompatibility complex (MHC) is located on the short arm of chromosome six at 6p21.31 (Fig. 1). The MHC was one of the first regions of the human genome that was completely sequenced and remains the most gene-dense region of the human genome identified and sequenced (The MHC sequencing consortium, 1999; Horton et al., 2004). The size and extent of the MHC has undergone a number of revisions from the early linkage maps (Carroll et al., 1987). Sequencing data (Horton et al., 2004) extended both the functional associations and size of the complex

HLA from Benchtop to Bedside
ISBN 978-0-12-823976-6
https://doi.org/10.1016/B978-0-12-823976-6.00002-0

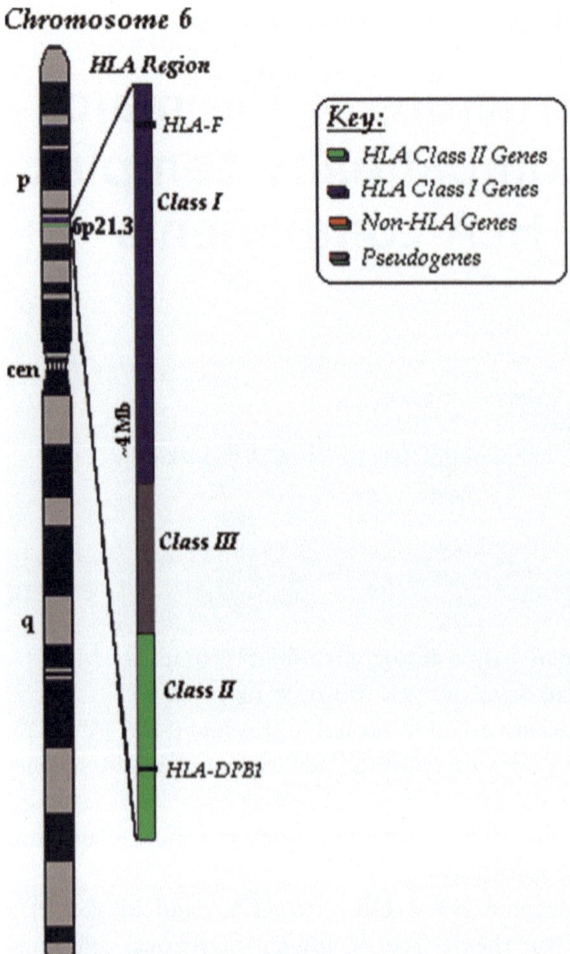

Figure 1 General structure of the MHC. *(From http://www.ncbi.nlm.nih.gov/projects/gv/mhc.)*

from around 3.5 megabases (Mb) to the more current understanding of approximately 7.6 Mb. This is important because the MHC also contains the most polymorphic human gene loci: Human leukocyte antigen (HLA) Class I and Class II (Mungall et al., 2003). Over 12000 HLA alleles were recognized by early 2014 (http://hla.alleles.org).

The HLA sections of the MHC are divided into three general regions called "classes" (Fig. 2). HLA Class I includes the first HLA locus identified, HLA-A (see Chapter 1), and two additional loci that were identified later, HLA-B and HLA-C (HLA-A, -B, and -C are frequently called the "classical" HLA loci). The MHC Class II region includes the complex

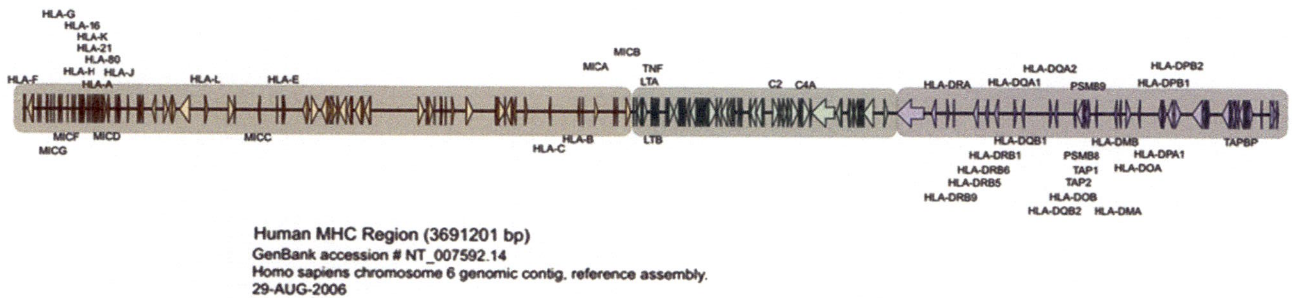

Figure 2 Map of the Human MHC Region. *(Used with permission, Ian A. York, Michigan State University.)*

HLA-DR, -DQ, and -DP loci, and the HLA Class III region contains numerous immune function related genes which are primarily expressed as plasma proteins.

HLA class I

HLA Class I proteins are dimers of a highly polymorphic "heavy" or alpha chain (45kD) and a monomorphic "light" or beta chain (12kD) similar to the structural arrangement of the evolutionarily related immunoglobulin molecules (HLA genes are part of the immunoglobulin superfamily). The heavy chains have their genes in the HLA Class I region, and the light chain is the highly conserved beta-2 microglobulin (Figs. 3A and 3B) for which the gene is located on chromosome 15 (15q21-q22.2). HLA Class I molecules are cell surface proteins (although secretory forms are found for some alleles). Structurally similar to other members of the immunoglobulin superfamily, HLA Class I proteins are built from a series of structural

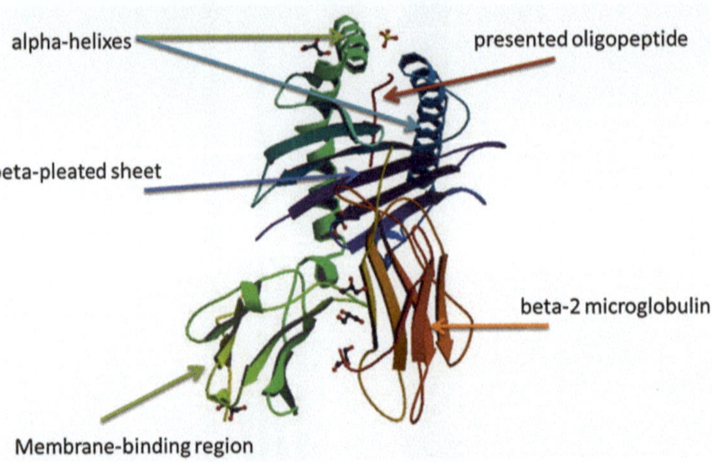

Figure 3A HLA Class I "ribbon diagram" (From HLA-A*02 _protein_structure.jpg, Wikimedia Commons). The red-orange structure to the lower right is beta-2 microglobulin. The membrane-binding region of the HLA Class I molecule is to the lower left. The "business end" of the molecule is the groove formed from the two alpha-helixes (the sides of the groove), and the beta-pleated sheet forms the base of the groove. Except for HLA-F, HLA Class I molecules are not expressed on the cell surface without an oligopeptide present in the groove (red linear structure approximately top center). *(Wikimedia image Madura, F., Rizkallah, P.J., Miles, K.M., Holland, C.J., Bulek, A.M., Fuller, A., Schauenburg, A.J.A., Miles, J.J., Liddy, N., Sami, M., Li, Y., Hossain, M., Baker, B.M., Jakobsen, B.K., Sewell, A.K., Cole, D.K., 2013. T-cell receptor specificity maintained by altered thermodynamics. J. Biol. Chem. 288 (26), 18766–18775.)*

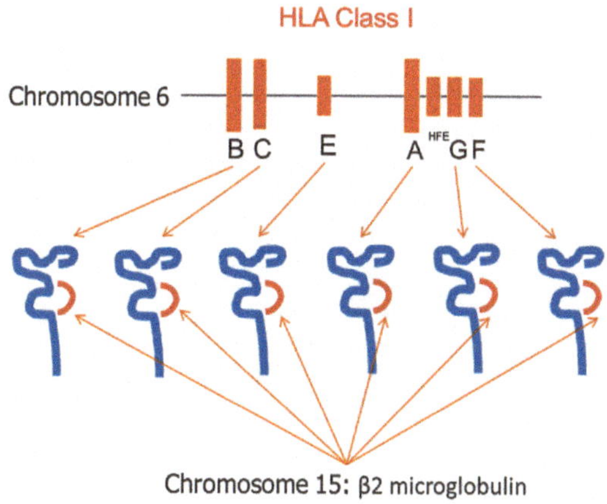

Figure 3B HLA Class I cartoon showing the "heavy chain" sources from chromosome 6 and the "light chain" source from the highly conserved β2 microglobulin gene on chromosome 15.

"domains" built from sequential exons (Malissen et al., 1982; López de Castro et al., 1982). Most of the antigenic and nucleotide polymorphism in the Class I heavy chain is located in the NH2-terminal domain which corresponds to the polypeptide binding groove (similar to the antigen-binding portion of immunoglobulins). Also similar to other members of the immunoglobulin superfamily, there are conserved extracellular domains which provide structure and ligand binding sites, followed by a trans-membrane region and a complex, anchoring, intracellular domain with the carboxy-terminus.

The other coding loci in the HLA Class I region produce proteins that are very similar to the "classical" HLA-A, -B, and -C products (The "classical" HLA loci are frequently referred to as class Ia). Unlike HLA-A, -B, and -C, however, the other HLA Class I molecules (Class Ib: HLA-E, HLA-F, HLA-G, HFE) have restricted tissue expression (Note: HLA-D, following the naming conventions from the International Workshops, was used for the "cellularly defined" antigens which became recognized as HLA Class II and covers the more complex HLA-DR, HLA-DQ, and HLA-DP loci).

Function of HLA Class Ia molecules

HLA Class I molecules present processed antigen to CD8 (cluster of differentiation 8) positive T lymphocytes through direct physical contact between the HLA protein and the T cell receptor (TCR) on the T cell. The complex interaction between the antigen-presenting cell (APC) and

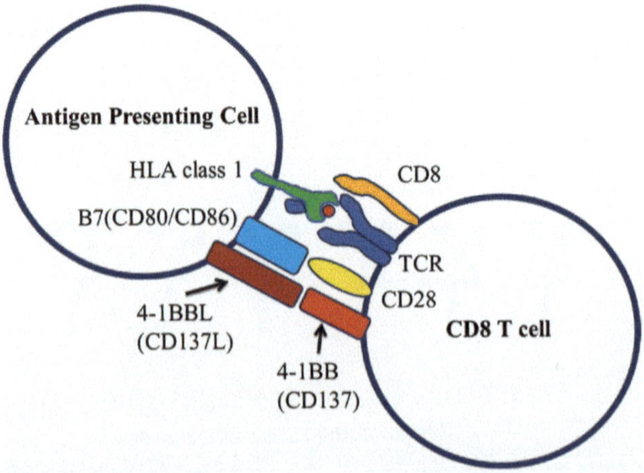

Figure 4 Costimulatory signals in CD8+ T cell interaction with antigen-presenting cell.

the T cell is referred to as the TCR-HLA (or MHC) complex. The CD8 molecule on the T cell acts as a stabilizer for the TCR-HLA interaction along with CD28 which binds its ligand B7 (CD80/86) on the APC and binding of 4-1BBL (CD137L) on the APC by 4-1BB (CD137) on the T cell (Fig. 4). CD8 positive T cells are identified as "cytotoxic T cells" and function by direct cell-to-cell killing of target cells that present nonself oligopeptides in the groove of their HLA Class I molecules. This allows the CD8 positive T cell to recognize cells that are infected by viruses or transformed by malignancy. Infected or transformed cells will have viral protein or cancer neoantigen peptides presented in the HLA molecule's oligopeptide binding groove which is formed from the two antiparallel alpha helices at the sides and an eight-stranded beta-pleated sheet as the bottom (Fig. 2.3A).

The antigen oligopeptides presented to CD8 T cells are from processed endogenous intracellular proteins ("self-antigens") which are manipulated by endoplasmic proteasome complexes and transported to the lumen of the endoplasmic reticulum (ER) by a transporter associated with antigen processing (TAP) protein complex (Fig. 5). TAP actively moves oligopeptides, within a limited size range, from the cytosol to the ER lumen. The endopeptidase activity in the cytosol and the proteasomes lyse (cut) proteins resulting in oligopeptide degradation products of varying lengths. The TAP complex includes large multifunctional proteases (LMPs; Belich et al., 1994) and is **selective** for peptides that are eight to 10 amino acids long.

The binding of the oligopeptide within the antigen-binding groove is dependent on the length, side chains, and charge distribution of the

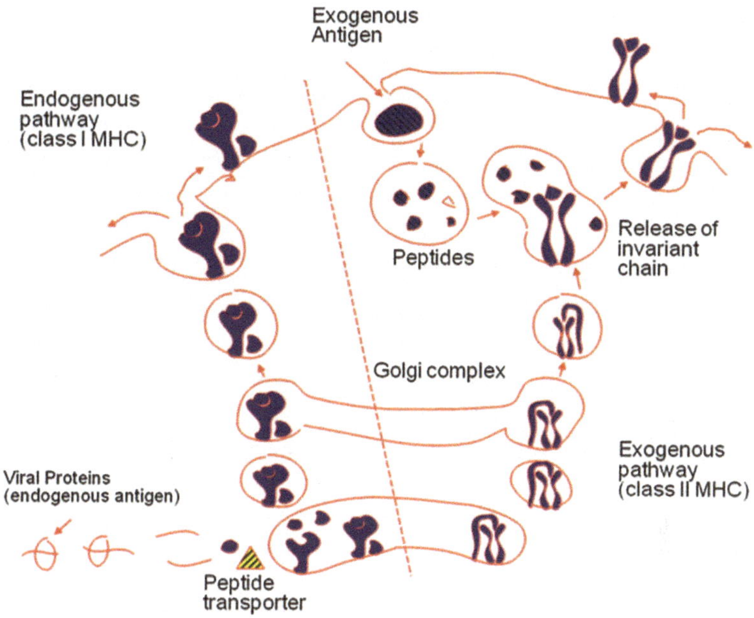

Figure 5 Antigen processing for HLA Class I and HLA Class II expression.

peptide. If you think of a "lock-and-key" relationship (the "lock" is the antigen-binding groove, and the "key" is the oligopeptide presented in the groove), you will have a good picture: some oligopeptides will have very stable binding, others will be readily replaced in the groove by an oligopeptide that better "matches" the size and shape of the groove while aligning with the charges of the amino acids that form the sides and bottom of the groove (the alpha helices that form the sides of the groove are most important). This process is called peptide editing.

Newly synthesized HLA Class I molecules are inserted into the inner membrane of the ER and stabilized by a molecular complex of transport proteins and "chaperones" that include TAP, tapasin, TAPBPR, calreticulin, ERp57, protein disulfide isomerase, and calnexin until beta-2 microglobulin binds, and an oligopeptide is inserted in the antigen-binding groove. Tapasin has been shown to be the "dedicated" Class I chaperone. The chaperone complex of tapasin, calreticulin, and ERp57 is called the peptide-loading complex. Binding of an oligopeptide to the antigen-binding groove with sufficient binding energy releases the chaperones. The "finished" HLA Class I molecule is transported to the cell surface by a secretory pathway (Kobayashi and van den Elsen, 2012). The editing process involves conformational changes in the heavy chain which "widen" the groove and depress the beta-pleated sheet until a peptide with higher binding energy displaces the chaperone. This process is repeated until a stable heavy chain, peptide and beta-2-microglobulin complex is formed (Jiang et al., 2017; Thomas and Tampé, 2017). HLA Class I molecules without oligopeptides in the groove or loosely bound peptides are degraded by ER-associated proteases and not expressed on the cell surface (with the exception of HLA-F).

The repertoire of oligopeptides presented by each HLA Class I molecule is limited by the size (length) selection performed by the proteosomal peptidases, the size (length) of the oligopeptide selected, and transported by TAP into the ER, and, finally, by the size (length) and fit in the peptide-binding groove which is limited by the charge and physical size of the amino acid side groups (Barber et al., 2001; Neefjes et al., 2011; Mommen et al., 2014). The final peptide is usually shortened to nine amino acids in length by ER aminopeptidase 1 (ERAP1) or ERAP2 (amino-terminal trimming). What the CD8 positive T cell "sees" is the presented oligopeptide in the specific context of the presenting HLA Class I molecule (Note that peptide lengths for presentation by MHC Class I tend toward nonapeptides but may vary between loci and alleles at individual loci (Rammensee, 1995)).

When the TCR of a CD8 positive T cell encounters an HLA Class I molecule, there are two likely results. If the combination of HLA molecule and presented antigen is one that the T cell "learned" to recognize as "self" during maturation in the thymus, there is no response, and the T cell continues on its way to "examine" other cells. If, however, the oligopeptide being presented is novel to the T cell (a "new" peptide being presented), the T cell becomes activated, releases cytokines that attract, and recruits other T cells. The T cell may release cytotoxins to directly kill the cell presenting the novel HLA/antigen combination. This process of examining or interrogating the HLA Class I molecules and the antigens they present is called immune surveillance, and is critical for the immune response to viral infections and malignant transformation of an individual's cells (The concept of immunological surveillance predates the characterization of the human MHC (Burnet, 1957; Thomas, 1959; Wilson, 1970)).

An additional function of the HLA class I molecules is as ligands for killer-cell immunoglobulin-like receptors (KIRs) (Parham, 2005). The KIR gene products are expressed on the surface of natural killer (NK) cells (innate immunity) which are lymphoid cells without the antigen restrictions (adaptive immunity) seen in helper (CD4 positive) and killer/suppressor

(CD8 positive) T lymphocytes. KIR genes are not part of the MHC and are located at chromosome 19q13.4. KIR/HLA interactions are important in recognition and destruction of virally infected cells and cancer by activating NK cells to kill the targets. Alternatively, KIR/HLA interactions, particularly with HLA-G, suppress the immune response to the trophoblast during pregnancy and may also have a role in stimulating the development of the critical blood flow patterns in the developing placenta.

HLA Ib molecules

HLA-E serves as an inhibitor for NK cells and has a very restricted number of nonapeptides it is able to display in the antigen-binding groove. Although HLA-E expression can be upregulated in most tissues in response to inflammation cosignals, its primary function appears to be protecting placental trophoblast tissue from maternal NK cells. Likewise, HLA-F appears to be a ligand for receptors on NK cells, minimally expressed on the surface of cells other than the placental trophoblast, but is inducible by inflammation. Unlike the other HLA Class I molecules, HLA-F expression is not dependent on the presence of an oligopeptide in the antigen-binding groove (Wainwright et al., 2000).

In contrast, HLA-G is expressed primarily on placental trophoblast tissues and appears to be inducible in other tissues (heart transplants), but serves the same NK-protective function (downregulation of the immune response through interactions with ILT2, ILT4, and KIR2DL4 receptors) for the pregnancy (trophoblast) conferred by HLA-E and HLA-F (and HLA-C) (Lazarte et al., 2017). HLA-G appears to be unique in having isoforms which result from posttranscriptional alternative splicing of the mRNA (Nilsson et al., 2014).

A locus formerly known as HLA-H is now known by its functional name, HFE, and serves as a regulatory protein for the transferrin receptor interaction with transferrin. The human hemochromatosis protein gene (HFE) was named for high Fe (iron). Mutations in the HFE gene result in the variable forms of hemochromatosis (Seckington and Powell, 2015).

Other HLA class I histocompatibility genes: MICA and MICB

MICA (MHC Class I-related chain A) and MICB (MHC Class I-related chain B) are highly polymorphic loci located between HLA-B and the

MHC Class III region. The immunoglobulin superfamily gene products are 383 amino acid peptides that function by binding the NKG2D activating receptor which is expressed on all NK cells and CD8 positive T cells. MICA and MICB are normally expressed on epithelial and endothelial cells and on activated CD4 positive T cells and activated B cells. Unlike the HLA molecules, MICA and MICB do not bind beta-2-microglobulin and do not bind polypeptides for recognition. The normal function of MICA and MICB appears to be recognition of altered self by NK cells (e.g., neoplastic changes). Antibodies to MIC antigens are associated with graft rejection, uncommon, but most commonly associated with retransplant (Tonnerre et al., 2013; Agüera-González et al., 2009).

HLA class II

Like HLA Class I molecules, HLA Class II molecules are dimers composed of a heavy (alpha) and "light" (beta) chain, although the difference between the molecular weights of the HLA Class II alpha and beta chains is much less (approximately 33kD and 29kD, respectively) than between the HLA Class I alpha chain and beta-2 microglobulin (45kD and 12kD, respectively) (Germain, 1994). HLA Class II molecules also present antigen to T cells but are paired with CD4 positive T cells (Figs. 6 and 7). Both the alpha and beta chain genes of the HLA Class II molecules are located in the Class II region of the human MHC (Fig. 2). Whereas HLA Class Ia molecules are expressed on all nucleated cells, HLA Class II molecules are generally restricted to specialized APC and inducible in other cell types. The principle human APCs are circulating monocytes and sessile dendritic macrophages.

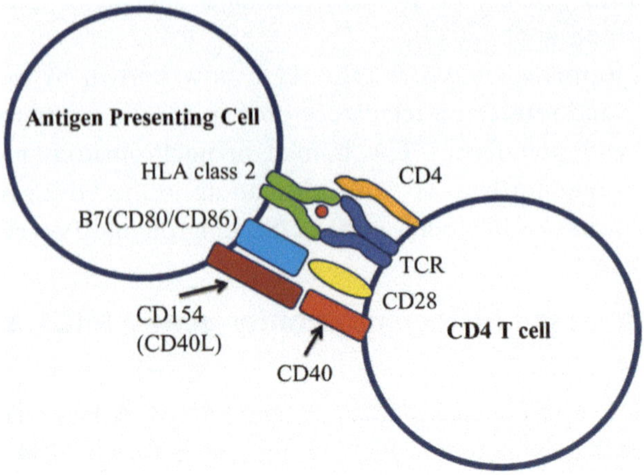

Figure 6 MHC Class II antigen presentation to T helper cells.

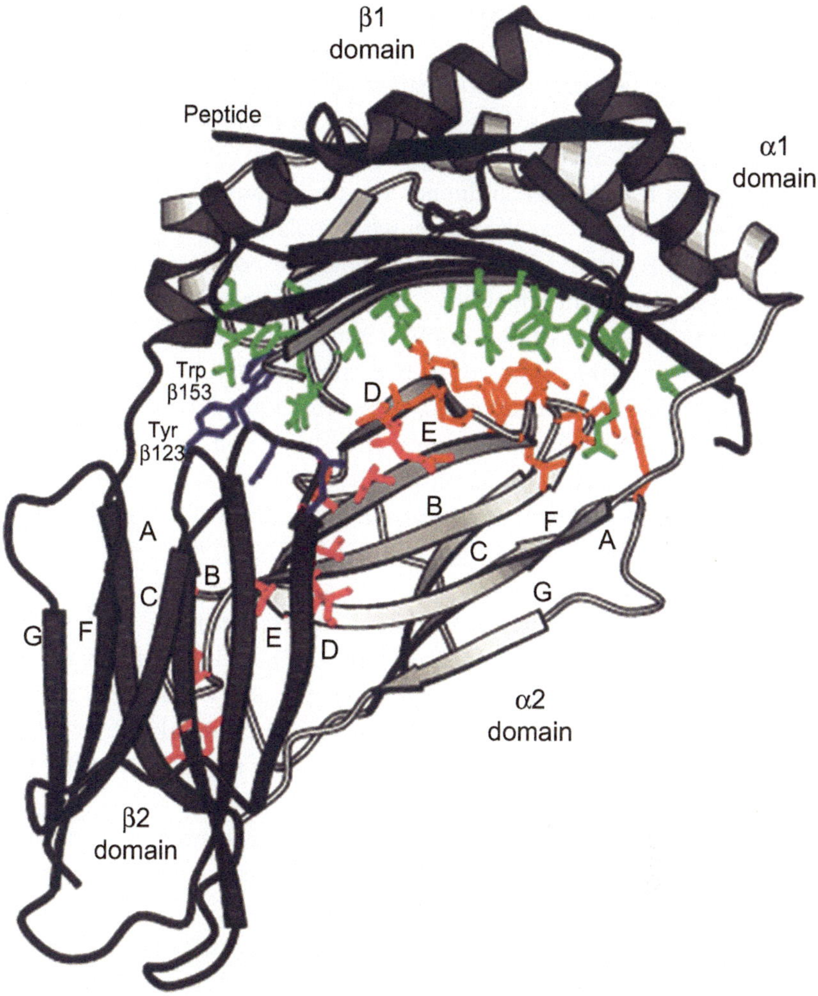

Figure 7 Ribbon diagram of an HLA Class II model showing the peptide-binding region domain interactions. A representative HLA-DRB1 heterodimer (HLA-DRA1/DRB1) with the beta chain (gray) closest to the viewer and the bound oligopeptide in the binding groove at the top. The HLA-DRA1 and -DRB1 constant regions are not shown. The colored residues indicate internal and interchain interactions within and between the alpha and beta chains *(see Murthy and Stern, 1997, Fig. 5 for details). Used with permission.* Note the similarity of the peptide binding groove structure to the HLA Class I molecule shown in Fig. 3A.

The HLA Class II loci were predicted as human analogs of the immune-associated (Ia) loci that had been characterized in the mouse H-2 system (Duquesnoy et al., 1979). Immune-associated Ia or Ia-like genes are involved in restricted antibody responses to foreign antigens (*Note that this use of "Ia" is different than the typographically identical HLA Class Ia which refers to HLA-A, -B, and -C loci*). In mice, the Class II or immune-associated loci predict whether or not a mouse strain will make antibody to foreign antigens such as bovine serum albumin or keyhole limpet hemocyanin (antigens to which mice are not normally exposed) (Lin et al., 1981).

HLA Class II genes also influence the proliferative response of T cells when they are exposed to self and nonself stimulating or target cells (usually lymphocytes) (Rich and Rich, 1974). These responses are measured by various combinations of target cells and responding lymphocytes as mixed lymphocyte cultures (MLC), primed lymphocyte cultures (PLC), and autologous MLC reactions. Once common in clinical laboratories supporting transplantation programs, these techniques are used primarily in specialized research today. The proliferative responses of the reacting lymphocytes were used to identify a major subset of "cell-mediated" immune responses and were used to characterize the HLA-D region (HLA Class II, now known to include the HLA-DR, -DQ, and -DP loci) (Shackelford et al., 1982).

Unlike HLA Class I, both the alpha and beta chains of HLA Class II molecules are polymorphic. For HLA-DR, -DQ, and -DP, the beta chain is much more polymorphic than the respective alpha chains. As a result, most of the variability that determines the "lock-and-key" fit of the presented oligopeptides into the binding groove of the Class II molecule is located in the amino terminal sequences of the beta chains.

HLA Class II molecules present external or environmental antigens to the CD4 positive T cells. The process (Fig. 5) begins with phagocytosis or pinocytosis of extracellular materials (bacteria, viruses, parasites, cell debris, etc) which are digested (broken down) within the phagolysosomal vacuoles of the APC by enzymes released during the merger of the phagosome with the lysosome. The MHC Class II protein is synthesized in the ER and stabilized in the lumen by the invariant chain (Ii) protein in the presence of calnexin. The MHC Class II/Ii complex is transported to the Golgi where endosomic microvesicles are budded off with the MHC Class II/Ii complex residing on the inside (not cytoplasmic side). The phagolysosomal vacuoles and the endosomes carrying the Class II molecules are merged, and the enzymes and acidic environment initiate lysis of the invariant chain leaving a fragment (Class II-associated invariant chain peptide or CLIP). A second MHC Class II protein, HLA-DM, facilitates "loading" of appropriately sized (length) peptides from the lysosomal complex and the removal of the CLIP (Sant and Miller, 1994). HLA-DM functions like tapasin in HLA

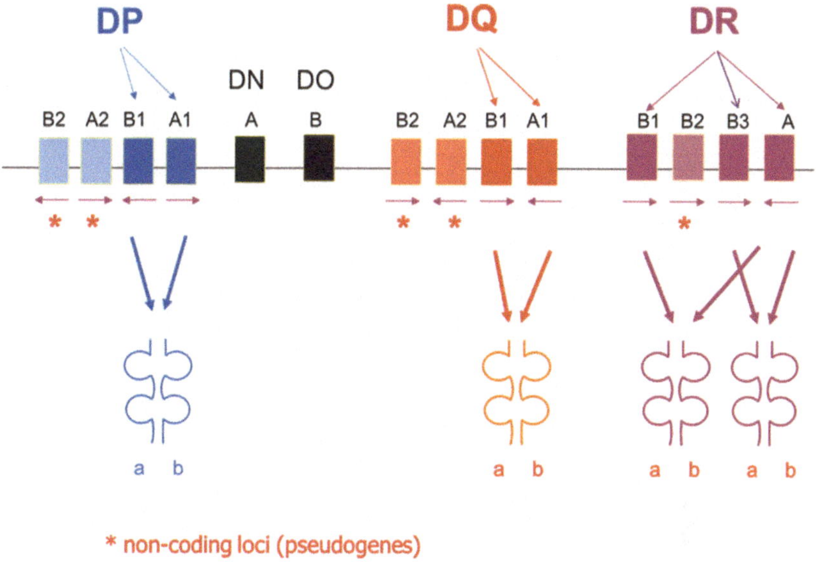

Figure 8 Cartoon of HLA Class II gene arrangements (centromere is to the left). Refer to Fig. 2 for more accurate relational map of the loci. The smaller lighter-colored blocks represent noncoding (pseudogene) loci associated with HLA-DR, HLA-DQ, and HLA-DP. Vertical arrows indicate which coding loci combine to form the HLA Class II surface antigens. Horizontal arrows indicate the relative sequence transcription directions for each locus.

Class I antigen peptide processing. The binding of peptides within the MHC Class II groove has been referred to as "promiscuous" because the binding is much less restrictive than that seen with MHC Class I molecules. For MHC Class II/peptide interactions, the relationship can be described as locks with alternate keys. Peptides eluted from MHC Class II binding sites range from 12 to 25 amino acids in length (in contrast to the 9 to 11 amino acid peptides eluted from MHC Class I molecules (Rammensee, 1995)).

Another significant difference between human MHC Class I and Class II genes is the existence of alternate beta chains available to combine with the alpha chain of the HLA-DR locus (Fig. 8). HLA-DRA provides the alpha chain for the beta chain from HLA-DRB1. Some HLA haplotypes (Fig. 9) carry an additional functional HLA-DR beta chain gene (either HLA-DRB3, HLA-DRB4, **or** HLA-DRB5). The maternal and paternal source alpha chains can combine with the beta chains that are cis (on the same chromosome) or trans (on the opposite chromosome). Likewise, the alpha chains can combine with the protein products of the other HLA-DR beta chains if present (HLA-DRB3 in the example shown in Fig. 8). The

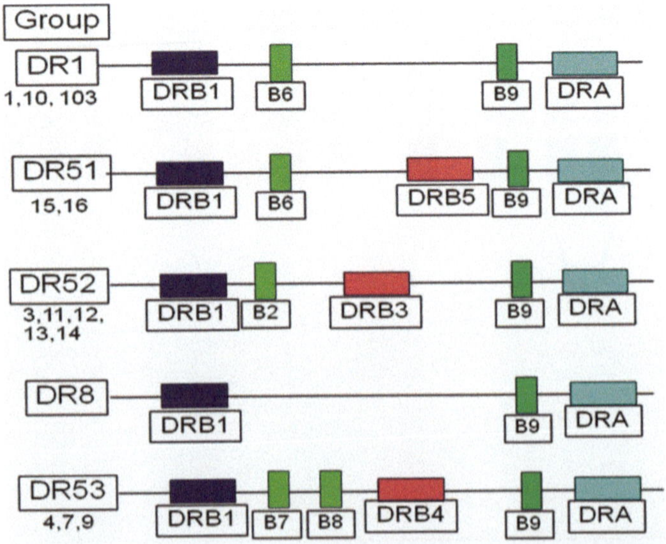

Figure 9 HLA-DR haplotypes. The smaller vertical boxes represent pseudogenes. The centromeric end is to the left in this diagrammatic representation. The haplotypes are named by the serological "group" with which each haplotype was first identified. The "numbers" under each group are the names of the HLA-DR antigens that are found in the group. *(After http://www.imgt.org/IMGTrepertoireMHC/LocusGenes/locus/human/MHC/hla_dr_representation.html.)*

alternate alpha/beta combinations increase the variety of oligopeptides that can be presented by the products of the HLA-DR locus on individual APCs. The combination of the "promiscuity" of the MHC Class II antigen presenting site and the alternative alpha/beta combinations of the HLA-DR locus significantly increase the "repertoire" of antigens that can be presented to the CD4 positive T cells.

In summary, HLA Class I molecules present intracellular self and modified self-antigens to CD8 positive T cells as part of immune surveillance for infection by viruses or neoplastic changes that could result in malignant tumors if unchecked. HLA Class II molecules present extracellular and environmental antigens to CD4 positive T cells which primarily stimulate B cell production of antibody specific for the presented antigens.

For additional reading about HLA presentation and the mechanisms for HLA Class I and HLA Class II antigen processing and the interactions between the peptides and binding grooves, I strongly recommend the review by Wieczorek et al. (2017). Two excellent, comprehensive, and accessible reviews of HLA structure and function are by Klein and Sato, 2000a,b, in the *New England Journal of Medicine (NEJM).*

References

Agüera-González, S., Boutet, P., Reyburn, H.T., Valés-Gómez, M., 2009. Brief residence at the plasma membrane of the MHC Class I-Related Chain B is due to clathrin-mediated cholesterol-dependent endocytosis and shedding. J. Immunol. 182, 4800–4808.

Barber, L.D., Howarth, M., Bowness, P., Elliott, T., 2001. The quantity of naturally processed peptides stably bound by HLA-A*0201 is significantly reduced in the absence of tapasin. Tissue Antigens 58, 363–368.

Belich, M.P., Glynne, R.J., Senger, G., Sheer, D., Trowsdale, J., 1994. Proteasome components with reciprocal expression to that of the MHC-encoded LMP proteins. Curr. Biol. 4 (9), 769–776.

Burnet, M.F., 1957. Cancer – a biological approach. 1. The process of control. Br. Med. J. 1, 779–782.

Carroll, M.C., Katzman, P., Alicot, E.M., Koller, B.H., Geraghty, D.E., Orr, H.T., Strominger, J.L., Spies, T., 1987. Linkage map of the human major histocompatibility complex including the tumor necrosis factor genes. Proc. Natl. Acad. Sci. USA. 84, 8535–8539.

Duquesnoy, R.J., Marrari, M., Annen, K., 1979. Identification of an HLA-DR associated system of B cell alloantigens. Transplant. Proc. 11, 1757–1760.

Germain, R.N., 1994. MHC-dependent antigen processing and peptide presentation: providing ligands for T lymphocyte activation. Cell 76, 287–299.

Horton, R., Wilming, L., Rand, V., Lovering, R.C., Bruford, E.A., Khodiyar, V.K., Lush, M.J., Povey, S., Talbot Jr., C.C., Wright, M.W., Wain, H.M., Trowsdale, J., Ziegler, A., Beck, S., 2004. Gene map of the extended human MHC. Nat. Rev. Genet. 5, 889–899.

Jiang, J., Natarajan, K., Boyd, L.F., Morozov, G.I., Mage, M.G., Margulies, D.H., 2017. Crystal structure of a TAPBPR-MHC I complex reveals the mechanism of peptide editing in antigen presentation. Science 358 (6366), 1064–1068.

Klein, J., Sato, A., 2000a. The HLA system. First of two parts. N. Engl. J. Med. 343 (10), 702–709.

Klein, J., Sato, A., 2000b. The HLA system. Second of two parts. N. Engl. J. Med. 343 (11), 782–786.

Kobayashi, K.S., van den Elsen, P.J., 2012. NLRC5: a key regulator of MHC class I-dependent immune responses. Nat. Rev. Immunol. 12, 813–820.

Lazarte, J., Adamson, M.B., Tumiati, L.C., Delgado, D.H., 2017. 10-year experience with HLA-G in heart transplantation. Hum. Immunol. 79, 587–593.

Lin, C.-C.S., Rosenthal, A.S., Passmore, H.C., Hansen, T.H., 1981. Selective loss of antigen-specific *Ir* gene function in *IA* mutant B6.C-$H\text{-}2^{bm12}$ is an antigen presenting cell defect. Proc. Natl. Acad. Sci. USA. 78 (10), 6406–6410.

López de Castro, J.A., Strominger, J.L., Strong, D.M., Orr, H.T., 1982. Structure of crossreactive human histocompatibility antigens HLA-A28 and HLA-A2: possible implications for the generation of HLA polymorphism. Proc. Natl. Acad. Sci. USA. 79 (12), 3813–3817.

Madura, F., Rizkallah, P.J., Miles, K.M., Holland, C.J., Bulek, A.M., Fuller, A., Schauenburg, A.J.A., Miles, J.J., Liddy, N., Sami, M., Li, Y., Hossain, M., Baker, B.M., Jakobsen, B.K., Sewell, A.K., Cole, D.K., 2013. T-cell receptor specificity maintained by altered thermodynamics. J. Biol. Chem. 288 (26), 18766–18775.

Malissen, M., Malissen, B., Jordan, B.R., 1982. Exon/intron organization and complete nucleotide sequence of an HLA gene. Proc. Natl. Acad. Sci. USA. 79 (3), 893–897.

Mommen, G.P.M., Frese, C.K., Meiring, H.D., van Gaans-van den Brink, J., de Jong, A.P.J.M., van Els CACM and Heck, J.R., 2014. Expanding the detectable HLA peptide repertoire using electron-transfer/higher-energy collision dissociation (EThcD). PNAS Early Edition. https://doi.org/10.1073/pnas.1321458111.PNAS.-March.10.2014. Published online before print March 10, 2014.

Mungall, A.J., Palmer, S.A., Sims, S.K., Edwards, C.A., Ashurst, J.L., Wilming, L., Jones, M.C., Horton, R., Hunt, S.E., Scott, C.E., Gilbert, J.G., Clamp, M.E., Bethel, G., Milne, S., Ainscough, R., Almeida, J.P., Ambrose, K.D., Andrews, T.D., Ashwell, R.I., Babbage, A.K., Bagguley, C.L., Bailey, J., Banerjee, R., Barker, D.J., Barlow, K.F., Bates, K., Beare, D.M., Beasley, H., Beasley, O., Bird, C.P., Blakey, S., Bray-Allen, S., Brook, J., Brown, A.J., Brown, J.Y., Burford, D.C., Burrill, W., Burton, J., Carder, C., Carter, N.P., Chapman, J.C., Clark, S.Y., Clark, G., Clee, C.M., Clegg, S., Cobley, V., Collier, R.E., Collins, J.E., Colman, L.K., Corby, N.R., Coville, G.J., Culley, K.M., Dhami, P., Davies, J., Dunn, M., Earthrowl, M.E., Ellington, A.E., Evans, K.A., Faulkner, L., Francis, M.D., Frankish, A., Frankland, J., French, L., Garner, P., Garnett, J., Ghori, M.J., Gilby, L.M., Gillson, C.J., Glithero, R.J., Grafham, D.V., Grant, M., Gribble, S., Griffiths, C., Griffiths, M., Hall, R., Halls, K.S., Hammond, S., Harley, J.L., Hart, E.A., Heath, P.D., Heathcott, R., Holmes, S.J., Howden, P.J., Howe, K.L., Howell, G.R., Huckle, E., Humphray, S.J., Humphries, M.D., Hunt, A.R., Johnson, C.M., Joy, A.A., Kay, M., Keenan, S.J., Kimberley, A.M., King, A., Laird, G.K., Langford, C., Lawlor, S., Leongamornlert, D.A., Leversha, M., Lloyd, C.R., Lloyd, D.M., Loveland, J.E., Lovell, J., Martin, S., Mashreghi-Mohammadi, M., Maslen, G.L., Matthews, L., McCann, O.T., McLaren, S.J., McLay, K., McMurray, A., Moore, M.J., Mullikin, J.C., Niblett, D., Nickerson, T., Novik, K.L., Oliver, K., Overton-Larty, E.K., Parker, A., Patel, R., Pearce, A.V., Peck, A.I., Phillimore, B., Phillips, S., Plumb, R.W., Porter, K.M., Ramsey, Y., Ranby, S.A., Rice, C.M., Ross, M.T., Searle, S.M., Sehra, H.K., Sheridan, E., Skuce, C.D., Smith, S., Smith, M., Spraggon, L., Squares, S.L., Steward, C.A., Sycamore, N., Tamlyn-Hall, G., Tester, J., Theaker, A.J., Thomas, D.W., Thorpe, A., Tracey, A., Tromans, A., Tubby, B., Wall, M., Wallis, J.M., West, A.P., White, S.S., Whitehead, S.L., Whittaker, H., Wild, A., Willey, D.J., Wilmer, T.E., Wood, J.M., Wray, P.W., Wyatt, J.C., Young, L., Younger, R.M., Bentley, D.R., Coulson, A., Durbin, R., Hubbard, T., Sulston, J.E., Dunham, I., Rogers, J., Beck, S., 2003. The DNA sequence and analysis of human chromosome 6. Nature 425, 805–811.

Murthy, V.L., Stern, L.J., 1997. The class II MHC protein HLA-DR1 in complex with an endogenous peptide: implications for the structural basis of the specificity of peptide binding. Structure 5, 1385–1396.

Neefjes, J., Jongsma, M.L.M., Paul, P., Bakke, O., 2011. Towards a systems understanding of MHC class I and MHC class II antigen presentation. Nat. Rev. Immunol. 11, 823–836 (Excellent review and highly recommended!).

Nilsson, L.L., Djurisic, S., Hviid, T.V.F., 2014. Controlling the immunological crosstalk during conception and pregnancy; HLA-G in reproduction. Front. Immunol. 5 (article 198), 1–10.

Parham, P., 2005. MHC class I molecules and KIRs in human history, health and survival. Nat. Rev. Immunol. 5, 201–214 (Excellent review).

Rammensee, H.-G., 1995. Chemistry of peptides associated with MHC class I and class II molecules. Curr. Opin. Immunol. 7, 85–96.

Rich, S.S., Rich, R.R., 1974. Regulatory mechanisms in cell-mediated immune responses. I. Regulation of mixed lymphocyte reactions by alloantigen-activated thymus-derived lymphocytes. J. Exp. Med. 140, 1588–1603.

Sant, A.J., Miller, J., 1994. MHC class II antigen processing; biology of the invariant chain. Curr. Opin. Immunol. 6, 57–63.

Seckington, R., Powell, L., September 17, 2015. HFE-associated hereditary hemochromatosis. 2000 apr 3. In: Pagon, R.A., Adam, M.P., Ardinger, H.H., et al. (Eds.), GeneReviews® [Internet]. University of Washington, Seattle, Seattle (WA), pp. 1993–2016. Available from: https://www.ncbi.nlm.nih.gov/books/NBK1440/. (Accessed 17 October 2016). link valid.

Shackelford, D.A., Kaufman, J.F., Korman, A.J., Strominger, J.L., 1982. HLA-DR antigens: structure, separation of subpopulations, gene cloning and function. Immunol. Rev. 66, 133–187.

The MHC sequencing consortium, 1999. Complete sequence and gene map of a human major histocompatibility complex. Nature 401, 921–923.

Thomas, C., Tampé, R., 2017. Structure of the TAPBPR-MHC I complex defines the mechanism of peptide loading and editing. Science 358 (6366), 1060–1064.

Thomas, L., 1959. In: Lawrence, H.S. (Ed.), Cellular and Humoral Aspects of Hypersensitivity. Hoeber-Harper, New York.

Tonnerre, P., Gérard, N., Chatelais, M., Poli, C., Allard, S., Cury, S., Bressollette, C., Cesbron-Gautier, A., Charreau, B., 2013. MICA variant promotes allosensitization after kidney transplantation. J. Am. Soc. Nephrol. 24, 954–966.

Wainwright, S.D., Biro, P.A., Holmes, C.H., 2000. HLA-F is a predominantly empty, intracellular, TAP-associated MHC class Ib protein with a restricted expression pattern. J. Immunol. 164, 319–328.

Wieczorek, M., Abualrous, E., Sticht, J., Álvaro-Benito, M., Stolzenberg, S., Noé, F., Freund, C., 2017. Major histocompatibility complex (MHC) class I and MHC class II proteins: conformational plasticity in antigen presentation. Front. Immunol. 8 (292), 1–16.

Wilson, D.B., 1970. Immunological surveillance. Science 169, 1006–1011. http://hla.alleles.org.

CHAPTER 3

Nomenclature and naming conventions for HLA

Contents

Bullet points

1. Antigen: A substance, usually a protein, to which an antibody can be made. Antibodies can be made to proteins, complex carbohydrates and other molecules which are, usually, foreign to the exposed human or other vertebrate. Antigens may also be defined by cellular reactions between a target cell and a responder. A cellular response may identify an antigen which is not recognized by antibody from the same species as well as antigens which are readily identified by antibody.
2. Epitope: The three-dimensional structure which is the binding site of an antibody that reacts with an antigen. An antigen may have multiple epitopes.
   a. Epitope group: A family of antigens which shares one or more epitopes that result in repeatable patterns of antibody-binding reactivity. Some epitopes are broadly shared between antigens and between loci. Some epitopes are very restricted and may define single antigen specificities.
   b. Epitope mapping: Analysis of the pattern of antibody binding across multiple groups of human leukocyte antigens (HLAs) and/or the analysis of shared amino acid sequences between HLAs that define relationships between antigens and alleles. Epitope mapping can be used to predict likelihood of antibody reactivity and rejection of transplanted organs and tissues.
3. Antibody: Immunoglobulin product of B cells and plasma cells, produced in response to an immunologic challenge and binds to the target antigen.

HLA from Benchtop to Bedside
ISBN 978-0-12-823976-6
https://doi.org/10.1016/B978-0-12-823976-6.00003-2

4. Cross-reacting: An antibody that recognizes epitopes on different antigens resulting in binding, usually of lower affinity or specificity, with families of antigens. The term cross-reacting can be used to refer to antibodies, antigens (i.e., cross-reacting antigen groups or CRAGS), or epitopes (cross-reacting epitope groups or CREGS).
5. Split: Antigen specificity that may be recognized from part of a broader pattern of reactivity. The definitions of many HLAs were divided (split) when reagent and testing methods demonstrated that a broad group of reactions revealed definable subgroups. For example, HLA-B17 was "split" into HLA-B57 and HLA-B58. The "ultimate" split is an allele.
6. Allele: Alternate nucleotide sequence at a gene. Alleles can differ by single nucleotides which may or may not change the amino acid coded for by the nucleotide triplet (codon) due to the degeneracy (redundancy) of the genetic code. Nucleotide changes that do not change the amino acid are referred to as "silent" and require sensitive molecular techniques for detection. Nucleotide changes that are present in more than 1% of the population are considered to be a polymorphism (allele). Most genes have a very common or "wild type" allele and rare allelic polymorphisms or mutations (less than 1%). There is no "wild type" for the human major histocompatibility complex (MHC) due to the massive polymorphism at the major loci.

The naming of human leukocyte antigens (HLAs) and genes is a complex moving target and was touched on in Chapter 1. Because of the dynamic nature of the nomenclature and the continually increasing number of recognized HLA alleles, the serious student of HLA is referred to the Nomenclature page of http://hla.alleles.org for the most current list of HLAs and alleles. The bullet points at the beginning of this chapter provide critical definitions which are necessary for understanding HLA nomenclature (Table 1).

The discovery of HLA antibodies preceded the identification of HLA, as described in Chapter 1. To summarize, the sera of many multiparous women were found to contain antibodies that would agglutinate the white blood cells of many, but not all, other people. The patterns of reactivity were determined to reflect the new antigen and genetic system now known as HLA and the major histocompatibility complex (MHC). The original naming of the antigens identified by the antibodies was determined by each laboratory or laboratories that identified a new pattern of reactivity. This lead to a confusing assemblage of names with no standardization outside of the original lab or group of labs. The successes of the international

Table 1 Antigen versus allelic complexity of HLA loci.

HLA locus	Number of recognized antigens	Number of recognized alleles
A	28	>5900
B	62 (including Bw4 and Bw6)	>7100
C	10	>5700
DR (DRB1 with DRA and DRB3, DRB4, and DRB5)	24	>3300
DQ (DQB1 and DQA1)	9	>1700
DP (DPB1 and DPA1)	6	>1500

From HLA numbers (http://hla.alleles.org/nomenclature/stats.html). Marsh, S.G.E, Albert, E.D., Bodmer, W.F., Bontrop, R.E., Dupont, B., Erlich, H.A., Fernández-Vina, M., Geraghty, D.E., Holdsworth, R., Hurley, C.K., Lau, M., Lee, K.W., Mach, B., Mayr, W.R., Maiers, M., Müller, C.R., Parham, P., Petersdorf, E.W., Sasazuki, T., Strominger, J.L., Svejgaard, A., Terasaki, P.I., Tiercy, J.M., Trowsdale, J., Nomenclature for factors of the HLA system, *Tissue Antigens* 75, 2010b, 291–455.

histocompatibility workshops that originated in Durham, North Carolina, with Dr. Bernard Amos chairing (1964), included the eventual acceptance and adoption of a standardized naming system that reflected the order of recognition of new antigens and the organization by genetic loci (Amos, 1968; Terasaki, 2007).

The international histocompatibility workshops were critical for establishing standardization of methods for antibody and antigen identification. As noted previously, HLAs were named by consensus of participants at international workshops in which cells and sera containing antibody were exchanged and methods of identification exchanged (Curtoni et al., 1967). Starting with the Fourth Histocompatibility Workshop, new antigens were given "workshop designations," if there was a consensus that a new antigen was being described and a "W" was placed in front of the next number in the sequence for the name of the antigen (e.g., W27) (Terasaki, 1970). When a "new" HLA was accepted, the workshop designation was removed. This naming convention worked for HLA-A and HLA-B, but did not carry over to HLA-C or to the other loci where numbering started over at 1 (one) for each new locus. As additional loci were added the workshop, "W" was changed to a lower case "w." HLA-C has the added naming complexity of retaining the use of the "w" before the antigen name (e.g., HLA-Cw7) to distinguish from the names of the complement proteins. Cw is used for HLA-C antigens 1 through 10.

NOTE: HLA and allele names contain numbers but are ordinal, not arithmetic. The naming convention became more complex and confusing as the immunologic relationship between antigens was elucidated, and initial attempts to maintain relationships between antigen names and allele names could not be maintained (Hurley et al., 1997). Naming became further complicated by the discovery of thousands of alleles at each HLA locus.

Initially, the naming of the HLAs was very closely linked to antibody-binding patterns. Cellular methods (e.g., mixed lymphocyte culture reactions) and molecular methods advanced the field and improved the ability to "type" for HLA. The methods for antigen and allele detection and identification will be discussed in a separate chapter. A very broad summary of the changes from the second half of the 20th century to the second decade of the 21st century is that new methods continue to require less patient sample and are both more sensitive and accurate. During 2 decades of service on the College of American Pathologists (CAP) Histocompatibility/Identity Testing Committee, I personally saw both the advancements and improved testing outcomes as demonstrated by the performance of HLA laboratories that participate in the CAP histocompatibility proficiency testing program.

Because HLA testing has its roots in serology, many terms that relate to the **naming** of HLAs come from serological testing and relate to the reactivity of the sera containing HLA antibody. Using HLA-A31 as an example (Fig. 10), the name of the antigen reflects evolution in the technology used to define the antigen and the reacting antibody. The name adopted after acceptance of the designation from the workshops was "A19" (19th recognized and segregated with the HLA-A locus). With improvements in techniques and evaluation of reactivity with many hundreds of

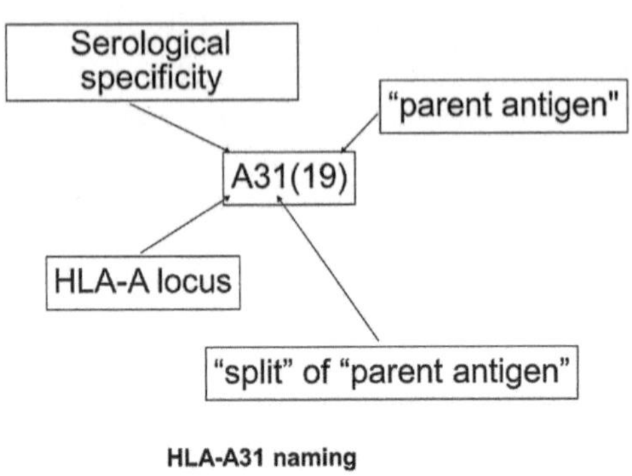

Figure 10 HLA naming using HLA-A31 as the example.

target cells, it was recognized that A19 could be "split" into new antigens (A29, A30, A31, A32, and A33). A19 is considered the "parent" antigen, and a serum that reacts with all of the A19 "family" is considered to have "broad" specificity. A similar pattern can be seen for HLA-B5 and the "splits" B51 and B52. There are multiple groups of HLAs that have been described in both HLA Class I and HLA Class II. Another pattern was seen with two specificities that were identified very early and named Bw4 and Bw6. Bw4 and Bw6 were found to be present with many HLAs and were called "broad" antigens. Subsequently, the Bw4 and Bw6 antigens were identified as epitopes found on multiple HLA Class I antigens and can be used to assist in differentiating related antigens (Vooter et al., 2000). So far, no antigen has been identified that has both the Bw4 and Bw6 epitopes.

The epitopes that define an antigen may be present on multiple antigens (e.g., Bw4 and Bw6) and are called "public epitopes." "Private" epitopes are distinct and are considered as specific or identifying for individual HLAs (Rody and Fuller, 1987). Based on the patterns of shared public and private epitopes and more recent sequencing data, evolution of the human MHC involved multiple gene duplications and extensive rearrangement between loci. As a result, epitopes may be shared between multiple antigens from one locus and between multiple antigens from different loci. An example of this complexity is the sharing of the Bw4 epitope across many HLA-B antigens and some members of three broad HLA-A antigen groups (http://hla.alleles.org/antigens/bw46.html).

HLA allele naming

The naming of HLA alleles has gone through multiple changes, driven by technology. The current nomenclature is cumbersome and confusing to both those who work with serology and molecular typing results. As noted previously, HLAs have been named in the order in which they were identified and accepted by the HLA research and clinical laboratory community. Likewise, HLA alleles have been named in order of identification and acceptance with the complication that the naming has been built on the preexisting nomenclature that was used for naming the antigens.

When molecular techniques were first introduced for HLA typing, the emphasis was on demonstrating that the nucleic acid sequences identified would give results consistent with the most reliable serological typings. As demonstrated by Dr. Gerhard Opelz (Mytilineos et al., 1990), molecular methods were able to reliably provide tissue typings and identified both mistypings and missed antigens. Very quickly, however, it became apparent that molecular methods were detecting "splits" of the splits detected by serology (subtypes) (Bodmer, 1988). Although some very well-defined HLAs like HLA-A2 were being shown to have subtypes by sensitive serological techniques and protein biochemistry (e.g., isoelectric focusing), more subtypes were detected by the newly introduced molecular methods. These subtypes were demonstrated to be alleles at each HLA locus and were numbered in order of discovery by adding two digits to the two digit antigen name (e.g., HLA-A*0201 and HLA-B*2701 and B*2702). The addition of the asterisk after the gene locus name was to conform to the conventions adopted by the Human Gene Mapping Nomenclature Committee (1979). Unfortunately, the number of alleles identified quickly exceeded the capacity of the four digit naming system.

The naming of HLA alleles has been complicated by the unprecedented polymorphism of the human MHC. The original naming conventions from 1987 were significantly modified during the 15th International Histocompatibility and Immunogenetics Workshop in Brazil, September 2008 (Marsh et al., 2010a, 2010b, 2010c). HLA-A*02 and HLA-B*15 both were found to have more than 100 alleles by 2002, and "rollover names" (A*92 for A*02 and B*95 for B*15) were soon found to be at capacity as new alleles were rapidly identified. As a result, colons (:) were introduced as delimiters into the names of the HLA alleles to accommodate the ever growing number of identified alleles and to provide a systemic structure for the components of the allele names.

The first section of the name of an HLA allele (Fig. 11) is the gene locus name (e.g., HLA-A). This is followed by the asterisk (*) to designate the allelic status (HLA-A*). The allele family name (e.g., 02 or 31, etc.) follows and the alleles within that family are delimitated by a colon (e.g., HLA-A*31:01). As previously discussed, the numbers are ordinal and are given in sequence of recognition. A second colon is added when there are allelic variants that do not change the amino acid sequence of the final protein, also called "silent variants" (e.g., HLA-A*31:01:02 as the second variant). The digits beyond the third colon (fourth group) indicate alleles in which the variation occurs in an intron or the untranslated sequences flanking the exons (e.g., HLA-A*31:01:02:03) (Marsh et al., 2010a, 2010b, 2010c; http://hla.alleles.org/nomenclature/naming).

It is important to note that the intention was for the allele family name (i.e., HLA-A*02) to correspond to the antigenic specificity, but it may not. There are at least two reasons. First, most HLA alleles are identified now by

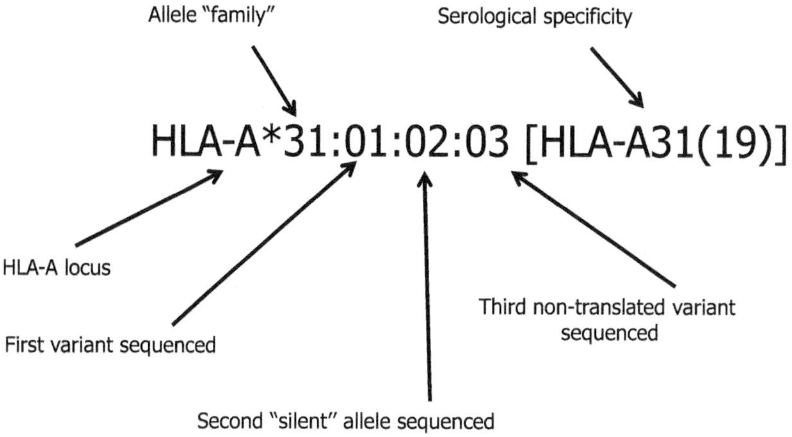

Figure 11 HLA allele naming using HLA-A*31:01:02:03 as the example.

molecular methods, and no serological studies may have been performed. Second, many HLA alleles are recombinations between alleles at a gene locus or recombinations between loci that result in novel combinations of serological "specificities." As a result, the WHO Nomenclature Committee for Factors of the HLA System, Buzios, Brazil, September 2008, wrote, "the allele name should be seen as no more than a unique designation" (Marsh et al., 2010a, 2010b, 2010c). The functional effect of the differences between the allele family name and the antigenic specificity will be discussed in the chapters on antibody detection and HLA typing.

Abnormally (or alternatively) expressed HLA alleles are designated by a capital letter following the allele name (e.g., HLA-A2*:02:101:01:02N). The aberrant expressions include null (N), low expression (L), secreted but not expressed on the cell surface (S), and questionable expression (Q; based on other alleles with similar nucleotide sequence changes). The cytoplasmic (C) and aberrant (A) classes of abnormal expression have not yet been assigned.

Two grouping designations (P and G) are provided by for use with ambiguous HLA allele typings (see Chapter 7) (Marsh et al., 2010a, 2010b, 2010c; http://hla.alleles.org/nomenclature/index.html). The P groupings (e.g., HLA-A*01:01P) are used to simplify reporting of HLA alleles that have the same antigen-binding domains. The P designation follows the 2 field allele designation of the lowest numbered allele in the group and will contain a minimum of four digits. The more frequently used G groupings (e.g., A*01:01:01G) simplify reporting of ambiguous typings for HLA alleles that share nucleotide sequences across the oligopeptide binding domains (exons 2 and 3 for HLA class I and exon 2 for HLA class II alleles). G group designations follow the first three fields of the allele designation for the lowest numbered allele in the group and must have at least six digits.

A "final" note about allele naming: there are gaps in the ordinal series! Named alleles may have been misidentified based on incomplete sequences or errors. When a named allele is found to have been identical to another named allele, or the sequencing fails confirmation, or another significant error is identified, the allele designation is removed from the list. For example, there is no HLA*B07:01. The list of deleted alleles designations (Marsh et al., 2010a, 2010b, 2010c) can be found at http://hla.alleles.org/alleles/deleted.html.

References

Alleles. http://hla.alleles.org/antigens/bw46.html.

Alleles. http://hla.alleles.org/alleles/deleted.html.

Amos, D.B., 1968. Human histocompatibility locus HL-A. Science 159, 659–660.

Bodmer, W.F. (reporter), 1988. Nomenclature for factors of the HLA system, 1987. Tissue Antigens 32 (4), 177–187.

Curtoni, E.S., Mattiuz, P.L., Tosi, R.M. (Eds.), 1967. Histocompatibility Testing 1967. Williams & Wilkins Company, Baltimore.

Hla. http://hla.alleles.org/nomenclature/naming.

Hla. http://hla.alleles.org/nomenclature/index.html.

Hurley, C.K., Schreuder, G.M.T., Marsh, G.E., Lau, M., Middleton, D., Noreen, H., 1997. The search for HLA-matched donors: a summary of HLA-A*, -B*, -DRB1/3/4/5* alleles and their association with serologically defined HLA-A, -B, -DR antigens. Tissue Antigens 50, 401–418.

Marsh, S.G.E., Albert, E.D., Bodmer, W.F., Bontrop, R.E., Dupont, B., Erlich, H.A., Fernández-Viña, M., Geraghty, D.E., Holdsworth, R., Hurley, C.K., Lau, M., Lee, K.W., Mach, B., Maiers, M., Mayr, W.R., Müller, C.R., Parham, P., Petersdorf, E.W., Sasazuki, T., Strominger, J.L., Svejgaard, A., Terasaki, P.I., Tiercy, J.M., Trowsdale, J., 2010a. An update to HLA Nomenclature. Bone Marrow Transplant. 45, 846–848.

Marsh, S.G.E., Albert, E.D., Bodmer, W.F., Bontrop, R.E., Dupont, B., Erlich, H.A., Fernández-Viña, M., Geraghty, D.E., Holdsworth, R., Hurley, C.K., Lau, M., Lee, K.W., Mach, B., Maiers, M., Mayr, W.R., Müller, C.R., Parham, P., Petersdorf, E.W., Sasazuki, T., Strominger, J.L., Svejgaard, A., Terasaki, P.I., Tiercy, J.M., Trowsdale, J., 2010b. Nomeclature for factors of the HLA system. Tissue Antigens 75, 291–455.

Marsh, S.G.E., Albert, E.D., Bodmer, W.F., Bontrop, R.E., Dupont, B., Erlich, H.A., Fernández-Viña, M., Geraghty, D.E., Holdsworth, R., Hurley, C.K., Lau, M., Lee, K.W., Mach, B., Maiers, M., Mayr, W.R., Müller, C.R., Parham, P., Petersdorf, E.W., Sasazuki, T., Strominger, J.L., Svejgaard, A., Terasaki, P.I., Tiercy, J.M., Trowsdale, J., 2010c. Nomeclature for factors of the HLA system. Tissue Antigens 75, 294.

Mytilineos, J., Scherer, S., Opelz, G., 1990. Comparison of RFLP-DR beta and serological HLA-DR typing in 1500 individuals. Transplantation 50 (5), 870–873.
Robinson, J., Halliwell, J.A., Hayhurst, J.H., Flicek, P., Parham, P., Marsh, S.G.E., 2015. The IPD and IMGT/HLA database: allele variant databases. Nucleic Acids Res. 43, D423–D431.
Rody, G.E., Fuller, T.C., 1987. Public epitopes and the antigenic structure of the HLA molecules. Crit. Rev. Immunol. 7 (3), 229–267.
Terasaki, P.I., 1970. Histocompatibility Testing 1970. Munksgaard, Copenhagen.
Terasaki, P.I., 2007. A brief history of HLA. Immunol. Res. 38, 139–148.
Vooter, C.E.M., van der Vlies, S., Kik, M., van den Berg-Loonen, E.M., 2000. Unexpected Bw4 and Bw6 reactivity in new alleles. Tissue Antigens 56, 363–370.

CHAPTER 4
The crossmatch

Contents

Bullet points:

1. A crossmatch is a test of compatibility between a potential recipient's serum and cells from a potential donor.
2. Human leukocyte antigen (HLA) antibody is most commonly formed after multiple pregnancies, organ transplant, and transfusion of cellular blood products containing leukocytes.
3. HLA antibody may be from any immunoglobulin class (IgM, IgG, IgA, IgE, IgD) or subclass (IgG3, 1, 2, 4, listed in order of complement-binding activity).
4. Complement-binding IgG is generally considered clinically relevant for transplantation.
5. Advances in HLA antibody detection have improved in both sensitivity and specificity.
6. Sensitivity of cell-based HLA antibody detection and crossmatching has increased with changes from agglutination to microcytotoxicity to flow cytometry.
7. HLA antibody may be formed to autoantigens and nonhuman histocompatibility antigens with crossreactivity to native human HLAs.
8. Lymphocyte-based crossmatch techniques do not detect antibodies to antigens that are not normally expressed on lymphocytes, such as MICA and MICB.

HLA from Benchtop to Bedside
ISBN 978-0-12-823976-6
https://doi.org/10.1016/B978-0-12-823976-6.00004-4

A **crossmatch** is performed to prevent antibody-mediated and rapid destruction of a transplanted organ, similar to the destruction of transfused red blood cells into an incompatible recipient (Bray, 2014). Although exceedingly rare today, hyperacute rejection, which occurs within minutes of transplantation and results in diffuse thrombotic destruction of the transplanted organ, was relatively common prior to the adoption of the pretransplant lymphocyte crossmatch (Terasaki, 2012). Today, most hyperacute rejection events are the result of clerical errors, including unintentional transplantation of blood group ABO-incompatible organs resulting in graft failure and occasional recipient deaths (Cecka, 2003). In addition to the routine use of the pretransplant crossmatch, many controls and procedural steps are routinely incorporated into the organ transplant process today to prevent these clinical disasters.

As noted in the first chapter, human leukocyte antigen (HLA) antibodies were discovered before the human major histocompatibility complex was identified. The earliest methods involved agglutination of peripheral blood leukocytes and distinguishing the patterns of agglutination and nonagglutination in both autoimmune and alloimmune situations (Dausset, 1962b). The target white blood cells used to determine presence or absence of antibody were initially prepared from mechanically defibrinated blood (rotation or agitation of whole blood with glass beads). Preparations of white blood cells from anticoagulated blood (ethylenediaminetetraacetic acid (EDTA or edetic acid) or sodium citrate) were quickly adopted due to the improved ease of recovery. Unfortunately, results obtained with cells from one method of isolation for leukoagglutination were often not reproducible with another isolation method, and reproducibility within isolation methods was less than 90% (Dausset, 1962a).

The naming conventions for the human immunoglobulins were not set until the 1964 World Health Organization (WHO) Naming Convention, Prague, CZ (Ceppellini et al., 1964). Most of the early publications referring to HLA antibodies discuss antibodies in terms of in vitro function or activity rather than the antibody class or subclass. As a result, it is common in older publications to find references to "partial" or "incomplete" antibodies when the (likely) antibody is IgG. This is because IgG is less able to crosslink and agglutinate cells compared to IgM, primarily due to the greater distance possible between the antigen-binding sites of the IgM molecule (approximately 1000 Ångstroms) relative to IgG (approximately 250 Ångstroms) (Widmann, 1989a). Although IgM is a pentameric antibody with 10 antigen-binding sites versus the two antigen binding sites on an IgG molecule, in agglutination each IgM molecule only crosslinks 2 two cells (lymphocytes or red blood cells, for example). IgM reactivity is more easily detected in the laboratory because the temperature optimum is closer to the room temperature of most laboratories. IgG usually is most reactive near 37°C.

Dausset (1962a) noted the association of development of "leukoagglutinins" with increasing numbers of blood transfusions and multiple pregnancies (Dausset, 1962b). This observation supported the hypothesis that the antibodies were formed in an immune response to "foreign" antigens not present in the person who made the antibody.

For the interest of those readers who are pursuing or practicing transfusion medicine, Dausset (1962a) noted that leukocyte-poor blood eliminated the transfusion reactions associated with leucoagglutinins and advised that leukocyte-poor blood be transfused to "any patient with or without agglutinins who present a reaction which cannot be accounted for" (Dausset, 1962b).

Microcytotoxicity and the "NIH standard" technique

Leukoagglutination methods consumed significant amounts of relatively rare serum and large amounts of peripheral blood for each test (Dausset, 1962b). Attempts to reduce reagent use brought the introduction of new methods like microcytotoxicity (Terasaki and McClelland, 1964) which became the standard of practice when later adopted by the United States National Institutes of Health (NIH) (Ray et al., 1976). This microcytotoxicity method became known as the "NIH Standard" and is still in use today although flow cytometric methods have largely replaced microcytotoxicity methods.

The microcytotoxicity method developed by Terasaki and McClelland resulted in a significant reduction in the volume of serum used in each test (1 microliter or one "lambda") per test reaction compared to 0.1 mL (Dausset, 1962b) and marked reduction in the number of cells needed for each reaction (200–250 cells per reaction well). Both changes conserved critical reagents and facilitated exchanges of cells and sera between different laboratories. The cell and serum exchanges allowed for rapid expansion of the knowledge of histocompatibility antigens and the antibodies made against those antigens after adoption of microcytotoxicity as the standard method for the Fourth Histocompatibility Workshop in 1970 (Terasaki, 2007). Isolation of mononuclear cells (primarily lymphocytes) for use in the microcytotoxicity assay was greatly improved by the use of a density gradient which separated neutrophils, erythrocytes, and platelets from the mononuclear cells with a single centrifugation step (Böyum, 1968).

Commercially prepared kits are readily available for isolation of lymphocytes from peripheral blood samples. Excellent step-by-step descriptions of isolation techniques can be readily found on the internet, from isolation kit manufacturers, and from professional societies such as the American Society for Histocompatibility and Immunogenetics (ASHI). A very reliable method is given by McCloskey et al. (1993). The authors also present reliable methods of isolation from lymph nodes and spleen.

The most basic lymphocyte microcytotoxicity test is incubation of isolated lymphocytes from peripheral blood (or lymph node or spleen from a deceased organ donor) with serum, usually at room temperature (Fig. 12). After the incubation, a source of complement is added (most frequently, diluted rabbit serum). After a second incubation, the cells are stained, usually with a fixative (initially, formaldehyde, but more recently, glutaraldehyde, due to lower toxicity for the person performing the test). Dye uptake or exclusion can be used to identify the killed versus unkilled cell populations. Complement fixation results in "holes" in the cell membrane created by the bound complement protein complex through which large dye molecules can pass. Chemical fixation stabilizes (preserves) the reaction so that it can be more conveniently interpreted (Figs. 13 and 14).

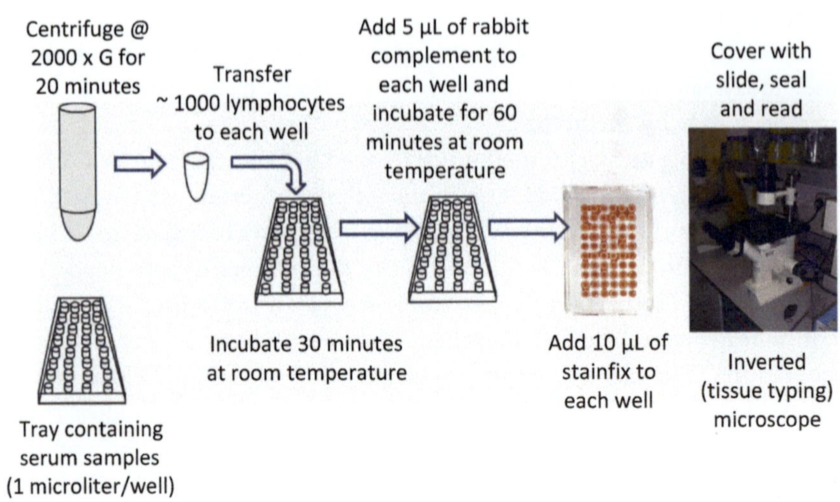

Figure 12 Lymphocyte microcytotoxicity assay.

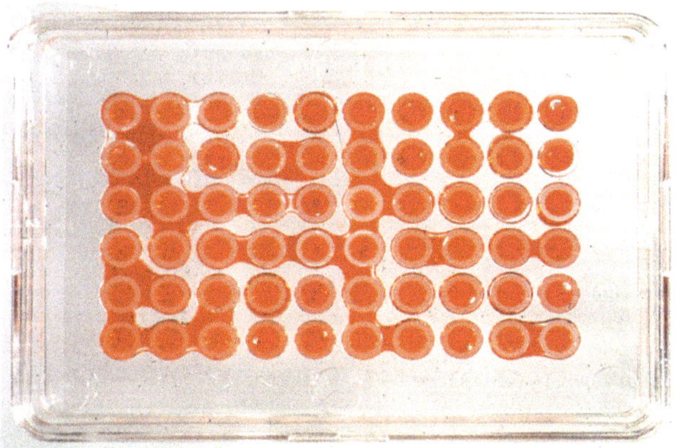

Figure 13 Lymphocyte microcytotoxicity tray with eosin-glutaraldehyde stain-fixative to reaction wells.

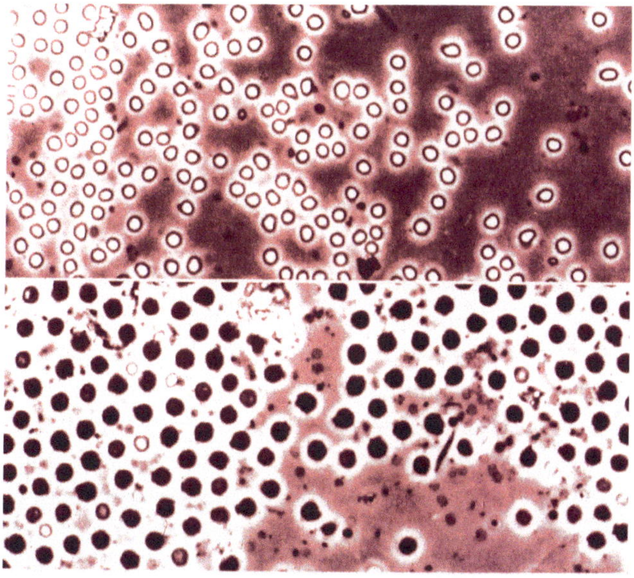

Figure 14 Lymphocyte microcytotoxity. The top field shows the transparent and highly refractile cells that have not been killed and stained in the reaction as viewed with an inverted phase contrast microscope (Fig. 13). Under the standard scoring practice (less than 10% killed cells), this would be scored "1" on the "0, 1, 2, 4, 6, 8" scale. The lower field shows the opaque, nonrefractile cells that have taken up the stain as evidence of complement-mediated cytotoxicity. Under the standard scoring practice, this would be scored "8" as more than 80% of the cells are dead.

There are many modifications of the "NIH Standard" technique of lymphocyte microcytotoxicity. A general statement about the modifications is that each is intended to improve sensitivity or specificity of the detection of binding of anti-HLA antibody to the target lymphocytes. Sensitivity may be increased by extending the incubation periods ("NIH extended") and specificity increased by wash steps (Amos et al., 1973).

Antihuman globulin

A radical change in the microcytotoxicity assay, with resulting increase in the sensitivity of detection of antibody, came with the introduction of an antihuman globulin (AHG) technique for HLA (Delmas-Marsalet et al., 1974). The antiglobulin technique was well established in immunohematology (Coombs et al., 1945) for the detection of weakly binding or low titer antibodies. Delmas-Marsalet et al. used the AHG technique to investigate cytotoxicity negative, absorption positive (CYNAP) reactions in which anti-HLA antibodies could be detected but were either too weak (low titer or concentration) to induce cytotoxicity or did not activate complement binding necessary for cell killing (e.g., IgG4). The AHG technique resulted in a significant (4- to 64-fold) increase in sensitivity for detection of anti-HLA antibody compared to agglutination and the NIH Standard microcytotoxicity methods (and non-AHG variants of that method) (Johnson, 1976a).

AHG reagent was initially prepared by immunizing rabbits with fractionated or whole human serum. The globulin fraction (enriched for immunoglobulins by ammonium sulfate or alcohol precipitation) generally gave the most acceptable and consistent reactions but has been replaced in current practice with antibody preparations that are specific for immunoglobulin subclasses such as IgG. The more broadly reactive "Coombs' reagents" (AHG) react with multiple immunoglobulin classes and complement proteins. Other species can be used to create an AHG reagent, but care should be taken to ensure that unintended cytotoxic and noncytotoxic antibodies to cell surface antigens are not present.

The AHG technique remains a standard method in laboratories performing lymphocyte microcytotoxicity. Although the AHG method is not specifically required, the use of a method that is more sensitive than the "NIH Standard" is enshrined in US federal regulations (42CFR493.1278(e)) for histocompatibility testing.

Crossmatching by microcytotoxicity

Microcytotoxicity has been nearly completely replaced by flow cytometric methods, but is included because the principles of crossmatching are consistent and the historical context is important.

The microcytotoxicity test is extremely robust and readily adapted to multiple purposes. When the lymphocytes of a potential organ donor are reacted with the serum of a potential organ recipient, this is a ***crossmatch***, similar to a red cell crossmatch in immunohematology. For living donor crossmatches, the target lymphocytes are isolated from peripheral blood. For deceased organ donor transplants, target lymphocytes can be isolated from peripheral blood, lymph nodes, or spleen for crossmatching. Crossmatches for living donors are usually between a single potential recipient and a single potential donor. For deceased organ donor crossmatches, there are usually multiple potential recipients for the kidneys, pancreas, heart, lungs, and liver of the donor. As a result, multiple donor to recipient combinations are usually set up for crossmatching at the same time, facilitated by the small volumes of serum, and relatively small numbers of donor cells needed to perform the crossmatch. For microcytotoxicity crossmatching of multiple organ *recipients* against a single donor, the donor cells are loaded into each well of the tray (Figs. 13 and 14) and serum from each potential recipient is added as in Fig. 12. (The trays may also be preloaded with sera from potential recipients to reduce time and was common practice in laboratories supporting large numbers of potential recipients (called the "waiting" or "wait" list). Current rules from the United Network for Organ Sharing (UNOS) and the Organ Procurement Transplantation Network (OPTN) require donor/recipient combinations (match run) be derived from a national database and single national computer system which negates the efficiency of preloaded trays). A positive control well with a pan-reactive serum and a negative control well using autologous serum from the donor or prescreened, antibody negative serum, are used to ensure that the reactions are valid. A standard 0, 1, 2, 4, 6, 8 scoring method is used. A positive reaction (greater than 10% cell killing; score of two or greater) is considered contraindicative for transplantation. Reproducibility is poor when less than 20% of the target cells are killed, so it is common practice to score crossmatches as negative when less than half of the replicates show more than 10% cell death (Johnson, 1976a).

ABO and solid organ transplantation

The A, B, and O blood groups are major markers of immunologic compatibility in transplantation, not just for blood transfusions. The A, B, and H antigens are broadly present on cell surfaces, including the endothelial linings of blood vessels. Unintentionally, ABO mismatched solid organ transplants are rapidly rejected (destroyed) by antibody-mediated complement deposition and thrombosis of the blood vessels. As with the pattern seen in red cell transfusion, individuals who are blood group O are considered "universal donors." Individuals who are blood group AB are "universal recipients," and blood groups A and B must receive organs from donors with their same blood group or blood group O (RhD is not present in transplant organs).

ABO blood groups are not evenly distributed in populations, and frequencies vary between racial groups. In the United States, 44% are group O, 42% group A, 10% group B, and 4% group AB. People with blood group B (and in some areas of the country, group O) waiting for organ donors are seen as "disadvantaged" because a smaller percentage of donors with blood group B are present than in the waiting recipient population. As a result, there have been many administrative and immunologic manipulations to improve transplantation, particularly of kidneys in the "immunologically disadvantaged" (Lipshutz et al., 2011).

T and B cell isolation

Sixty-five to eighty-five percent of peripheral blood lymphocytes are T cells in normal individuals (Williamson, 1993). When unfractionated cell populations from peripheral blood lymphocyte isolations are used in HLA crossmatching (or serological typing), the results reflect the predominantly T cell component. HLA Class II antigens, as previously discussed, are not normally expressed by T cells, and antibodies to HLA Class II antigens (primarily HLA-DR and −DQ) are unlikely to be detected by microcytotoxicity without isolating or enriching for B cells.

There have been many approaches to isolating or enriching B cell populations, including nylon wool (B cells adhere, T cells do not), sheep red blood cell (SRBC) rosetting. T cells spontaneously rosette with the SRBCs through T cell CD2 binding of the T11TS ligand on the SRBC. The rosettes can be removed by density-gradient isolation similar to the method of isolating lymphocytes from peripheral blood, and positive selection using antibody to B cell CD19 linked to magnetic beads (Brown et al., 1993).

Flow cytometric methods have eliminated the need for isolation of T and B cell fractions for most clinical assays in histocompatibility.

B cells are more difficult to work with than T cells for multiple reasons: lower numbers in peripheral blood relative to T cells, more complicated isolation, lower recovery fractions than T cells, and both lower initial viability and viability during preservation. The differences are intrinsic to the lymphocyte populations (Mallone et al., 2010). The challenges of working with isolated B cells also include the lower density of the HLA Class II antigens on the cell surface relative to Class I (Bux et al., 1993). The lower density of HLA Class II antigens results in added difficulty in detecting HLA Class II antibody in crossmatching, antibody screening, and antibody identification by microcytotoxicity. The reader is reminded that T lymphocytes do not normally express HLA Class II antigens and that B lymphocytes express both HLA Class I and Class II antigens.

Prior to the introduction of antibody-coated magnetic beads for isolation of lymphocyte subpopulations, an alternative was the use of antibody-coated ox erythrocytes (Moretta et al., 1976). Receptor-negative cells would not bind to the coated red cells (negative selection), and positive cells could be recovered by lysis of the ox erythrocytes.

Immunofluorescence

Antibody to HLAs can be detected by antihuman immunoglobulins conjugated to fluorescent dyes. The method can be very sensitive and does not rely on cytotoxicity but does require specialized microscopy or instrumentation. Slide and well techniques for immunofluorescence (IF) have been widely adopted in both research and clinical laboratory science for many analytes (Perkins et al., 1972).

Advantages of IF, in addition to the increased sensitivity, include the ability to use additional dyes to identify subsets of cell markers, subpopulations of cell types, and evaluation of cell viability (dye uptake and/or dye exclusion: Jones and Senft, 1985). Combining antibody specificity with distinct fluorescent dyes has allowed detection of antigens that are present only in specific subsets of cells, for example B lymphocytes, in the presence of other cell populations. For example, an antibody covalently linked ("labeled") with rhodamine isothiocyanate (RD) with specificity to a B cell specific antigen such as the cluster of differentiation (CD) antigen CD19 can be used to identify B lymphocytes, and a second antibody labeled with fluorescein isothiocyanate (FITC) with specificity to human

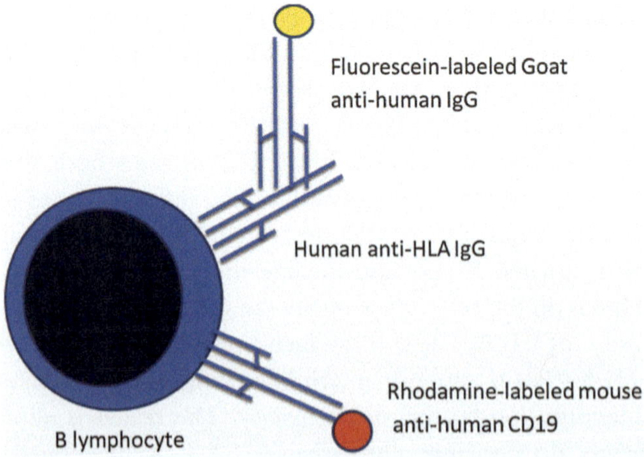

Figure 15 Cartoon of two color immunofluorescent staining of a B lymphocyte for bound antihuman HLA. The "primary antibodies" are the mouse antihuman CD19 antibody that identifies the B lymphocyte and the human anti-HLA IgG molecule. The "secondary antibody" is the fluorescein-labeled goat antihuman IgG that identifies human IgG that is binding to the cell surface.

immunoglobulins can identify the presence of the (predominantly) anti-HLA antibodies. In practice, the labeled antibodies are frequently specific for species-specific immunoglobulins such as a goat antihuman IgG or goat antimouse IgG to give an enhancement effect to the reactions similar to the AHG technique in microcytotoxicity (Fig. 15).

Disadvantages of IF include complicated and frequently expensive microscopy with a source of light to excite the fluorescent dye (epifluorescence) and filters that allow the photons from the excited dye to reach the observer while excluding the excitation wavelength. Compensation for autofluorescence of the cells being observed is obtained by comparison with negative controls. Another consistent limitation of fluorescent techniques is the short time that the fluorescent dyes remain active when exposed to light (bleaching effect) which also requires most fluorescent microscopy to be performed in darkened rooms. Although many methods have been introduced to stabilize the fluorescence, most IF methods remain temporally short-lived. Although once widely used in clinical laboratory medicine, fluorescent microscopy is being supplanted by other sensitive methods that give more stable reactions and are less subjectively evaluated.

Flow cytometry

Originally called "pulse cytophotometry," flow cytometric methods use a laser or other intense monochromatic light source (e.g., mercury arc lamp)

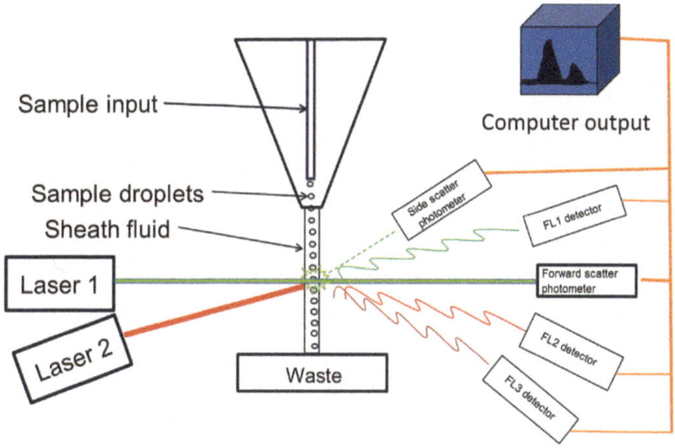

Figure 16 Flow cytometer schematic. Additional laser light sources allow for detection of additional fluorochromes.

to interrogate and analyze cell populations that have been treated with dyes, lectins, or antibodies with specificities for nuclear, cytoplasmic, or membrane targets. Like fluorescence microscopy, flow cytometric methods rely on detection of the energy released from excitation of electrons in fluorochromes covalently attached to antibodies (in most cases) to cell target antigens or immunoglobulins, such as anti-HLA antibodies (Fig. 15), bound to the surface of the cell. Modern flow cytometers bear little resemblance to the original instrument constructed in 1968 by Wolfgang Göhde in Germany (Kalodimou, 2013). The first flow cytometer I worked with took up most of a research laboratory (Butterfield et al., 1982). Most modern flow cytometers are relatively compact clinical instruments and widely adopted in clinical and research laboratories. Unlike fluorescence microscopy, flow cytometric methods suspend the target cells to be analyzed in a fluid column carried within a rapidly moving column of fluid called "sheath fluid" past the light source used to induce the fluorescent emission(s) (Fig. 16). An advantage over fluorescence microscopy methods is that the light emissions can be readily and objectively quantified by flow cytometers. Until relatively recently, light emissions in fluorescence microscopy were evaluated qualitatively and with subjective scoring.

Flow cytometers have replaced the human eye for detection of the fluorescent signal and phenotypic categorization of cell populations for many applications. A flow cytometer is composed of a sampling system that inputs the target sample into the stream of the sheath fluid, the monochromatic light source or sources, and photomultiplier tubes which are optically arranged to detect selected wavelengths of light emitted from the excitation of fluorescent dyes and the light that is refracted or reflected from the target cells (or other particles) carried in the sheath fluid (Fig. 16).

Relative size of the individual cells is obtained from ***forward scatter*** (usually abbreviated as FSC or FS) determined from the light that is refracted toward the detector/photomultiplier in line with the primary laser. The light from the laser is normally blocked by an "obscurator bar" fixed in line with the light source. Light that is refracted by a cell passing through the beam is scattered in the forward direction in proportion to the size (volume) of the cell. The larger the cell, the more the light is refracted "around" the "shadow" of the obscurator.

Cellular complexity (granularity and nuclear lobation, for example) is obtained from low angle ***side scatter*** (usually abbreviated as SSC or SS). Unlike FS, no obscurator is needed, and the scattered light is detected by a photomultiplier placed at an angle to the line of the primary light source.

The combination of FS and SS can be used to group blood cells into populations that correlate well with hematologic morphology (Figs. 17 and 18). Computer software is used to select cell populations for analysis ("gating"). For histocompatibility testing, the gated population is usually the small, nongranular population that corresponds to lymphocytes. The parameters of FS and SS can be combined with other cellular parameters to construct automated differentials of white blood cells and, when combined

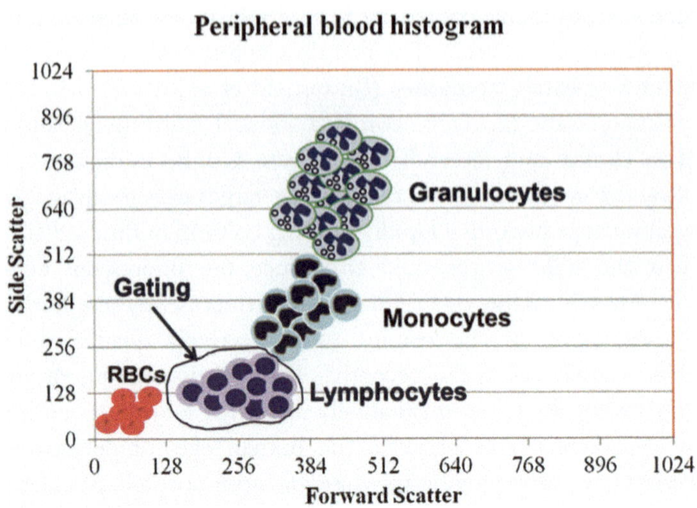

Figure 17 Light scatter histogram diagram for normal peripheral blood. Gating (indicated by a line around a phenotypic group) is obtained by computer-aided delimitation of the cell population selected for analysis.

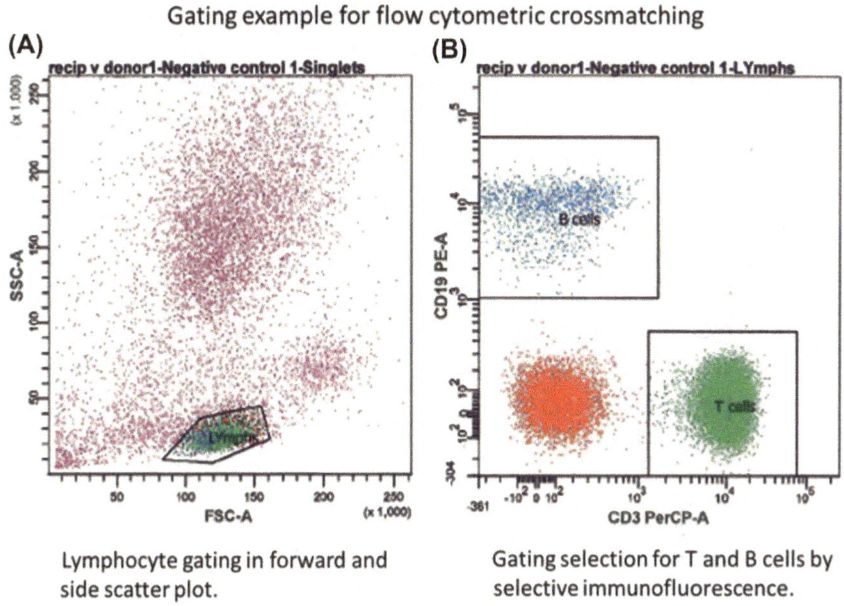

Figure 18 Computer-assisted gating for analysis. (A) Shows the gating of the lymphoid population by forward and side scatter properties. (B) Shows the gating for T cells (CD3 positive) by peridinin chlorophyll protein complex (PerCP) fluorescence and B cells (CD19 positive) by phycoerythrin (PE) fluorescence. (1024 channel output, FACSCanto II™, Becton, Dickinson and Company, Franklin Lakes, NJ).

with vital dyes and impedence measurements, are used in many modern clinical hematology analyzers.

Although histocompatibility testing is normally performed on lymphocyte samples that have been isolated by density gradient sedimentation or magnetic bead isolation, light scatter gating should be used to prevent inadvertent analysis of other nonlymphocyte cell populations and debris. Monocytes and neutrophils nonspecifically bind immunoglobulins through Fc-receptors, and false-positive binding of antibody can be expected (Maeda et al., 1996). Nonspecific binding of antibody may also be present due to B cell Fc receptor binding which can be reduced by pretreatment of the lymphocyte preparation with the proteolytic enzyme Pronase (Pronase E, protease from *Streptomyces griseus* Type XIV, Sigma Aldrich, St. Louis, MO) (Vaidya et al., 2001). Additional gating can be accomplished by analyzing only cells marked as T cells by antibodies to the pan-T marker CD3 labeled with peridinin chlorophyll protein complex (PerCP) and B cells using the B cell marker CD19 (or CD20) labeled with phycoerythrin.

Flow cytometric crossmatching ("**flow crossmatch**") is used to provide a more sensitive crossmatch test (compared to a microcytotoxicity crossmatch) and has largely replaced the AHG crossmatch in clinical use (unpublished 2014 data from MX-1 and MX-2 Proficiency Surveys, College of American Pathologists Histocompatibility/Identity Testing Committee, used with permission). The use of a flow cytometric crossmatch was first described in 1983 (Garovoy et al., 1983). The Garovoy paper compared extended incubation, Amos wash (Amos et al., 1969), and AHG microcytotoxicity crossmatching to flow crossmatching using a simple two-color method for T and B cell crossmatches. In this procedure, Garovoy et al. used rhodamine conjugated goat $F(ab')_2$[1] antihuman immunoglobulin to detect antibody binding to the surface of the density-gradient isolated lymphocytes. T cells were specifically identified using mouse monoclonal antibodies to the pan-T marker Leu-4 (now known as CD3) conjugated with fluorescein. Detection of HLA antibody was determined by the shift of the rhodamine fluorescence histogram peaks "to the right" (increased fluorescence intensity) by $\geq$9 channels on a 256 channel display of logarithmic data.

The photomultipliers in most flow cytometers produce continuous analog signals which are amplified and processed by analog-to-digital converters (ADCs). The digital output is displayed and analyzed on a computer. The output can be displayed on a linear or logarithmic scale. The processing capability of the ADC determines the number of channels that can be displayed. An 8-bit ADC can produce 2^8 channels (256 channels). More commonly today, a 10-bit ADC gives 2^{10} channels (1024 channels). Fluorescence intensity is usually analyzed with linear amplification. For a 10-bit ADC, an increase of 256 channels corresponds to a ten-fold increase in fluorescent intensity when displayed on a log scale. Note that each "channel" is actually the output of a range of electrical energies from the photomultiplier "lumped" together or "binned" to simplify the output for analysis. Whether output is linear or log-linear, channel shift is a proportional change.

Antibody binding in flow cytometry is usually measured as increased fluorescence intensity compared to a control. The control is usually target cells to which autologous or prescreened serum that is known to be

[1] $F(ab')_2$ antibody has the Fc portion of the immunoglobulin removed by pepsin proteolytic digestion to prevent nonspecific binding to Fc receptors on lymphoid cells. The Fc portion is readily crystallizable since the amino acid sequence variability is in the antigen binding (Fab) portion of the antibody molecule.

negative for anti-HLA antibody is added prior to the addition of the fluorochrome-labeled antihuman immunoglobulin. Comparison to control cells allows for assay "correction" for the inherent autofluorescence of cells and nonspecific binding of reagents to cell surface proteins. Each laboratory must determine the increase in (specific) fluorescence that is considered significant. This is usually reported as a "channel shift" or other measure of fluorescence intensity change (usually increased, Fig. 19A).

With linear amplification and display, increases in number of channels result in arithmetic changes (doubling the fluorescence doubles the number of channels). For logarithmic displays on a 10-bit ADC instrument with 1024 channels, doubling the number of channels corresponds to a ten-fold increase in fluorescence. Changes at low channel numbers represent much smaller changes in the amount of antibody bound than changes at higher channel numbers. (Again, note that channel shifts indicate proportional changes.)

HLA antibody detection in tissues

Reactivity of human antibody to HLAs in tissues has been documented by various techniques for over 40 years (Sybesma et al., 1974). The most common method has been IF. Antibody binding to the tissue is detected by a fluorescein-labeled secondary antibody with specificity to human immunoglobulins or is augmented by using a fluorochrome-labeled third antibody with specificity to the species from which the second antibody was prepared (Fig. 15).

Perhaps not obvious, using tissue for crossmatching is not clinically useful, so this is primarily used in surgical pathology to detect antibody as evidence of rejection after a solid organ transplant. The distribution of HLAs in tissues has been well characterized (von Willebrand et al., 1985) which allows correlation of antibody or complement deposition with the pathologic changes found in routine histological examinations. Grading criteria for pathologic changes in transplanted tissues are called the **Banff** schema and are regularly updated (Haas et al., 2014).

HLA antibody binding to tissue is detectable using IF in frozen tissue sections or immunohistochemistry (IHC) in frozen and formalin-fixed tissue sections. As previously described, IF uses a fluorophore-tagged primary or secondary antibody to signal antibody binding. IHC uses enzyme-tagged primary or secondary antibodies to generate color in tissue sections at the site of antibody binding (Falini and Taylor, 1983; Giorno, 1984).

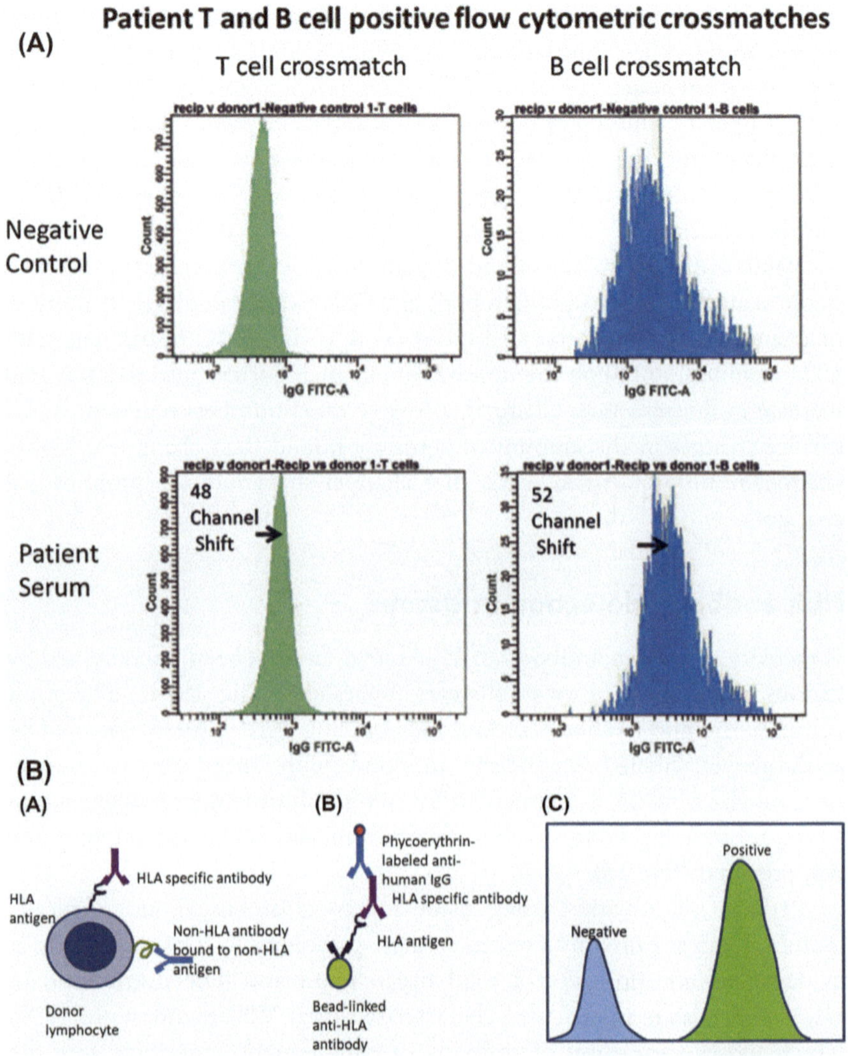

Figure 19 (A) Flow cytometric crossmatch demonstrating channel shifts seen with weakly positive HLA antibody reactivity. The patient was shown to have an antibody against a donor-specific HLA (DSA) (1024 channel output, FACSCanto II™, Becton, Dickinson and Company, Franklin Lakes, NJ). (B) Cartoon of donor-specific antibody bead-linked flow crossmatch (DSA-FXM). Panel A: postincubation specific and nonspecific antibody binding to HLAs and non-HLAs on donor lymphocytes. Panel B: postlysis binding of fluorescent bead linked to nonantigen binding region of HLA and phycoerythrin-labeled antihuman IgG. Panel C: representative flow histogram showing relative positions of positive and negative peaks. *(After Chen, G., Lin, L., Tyan, D.B., 2020. DSA-FXM: accelerated donor-specific flow crossmatch discriminating class I and II antibody specifically and only to donor HLA for determining true incompatibility. Transplantation 104(4) 813–822.)*

Common chromogenic enzymes are horseradish peroxidase (HRP), alkaline phosphatase (AP), and combinations with strepavidin and biotin (all require blocking of endogenous activity to obtain reliable results). As with any laboratory procedure, it is critical that manufacturer (vendor) instructions are followed, and the method is validated in the setting in which it is used.

Although anti-HLA antibody can be detected, in practice, it is the evidence of antibody-mediated complement fixation that is usually detected in clinical testing. Most commonly used, complement degradation fraction 4d (C4d) was identified in 1993 as a stable marker of donor-specific antibody binding in transplanted tissue (Feucht et al., 1993). C4d is covalently bound to endothelium after complement activation and fixation by complement-binding antibody (primarily IgM and IgG3 and IgG1; IgG2 binds complement poorly, and IgG4 is generally considered as not complement-binding) (Widmann, 1989a). In current clinical practice, detection of C4d deposition, histologic evidence of acute tissue injury (microvascular inflammation, arteritis, thrombotic microangiopathy, or acute tubular injury), and serologic evidence of donor-specific antibody to HLA or other antigens are required for diagnosis of acute or active antibody-mediated rejection in renal transplants (Haas et al., 2014). Other markers of rejection, including other complement components, are and have been utilized in the tissue diagnosis of rejection but are beyond the scope of this text (see Kooijmans et al., 1996; Regele et al., 2001; Kuypers et al., 2003; Djamali et al., 2014).

A novel solid-phase crossmatch technique has been described using detergent-solubilized preparations of HLAs from donor lymphocytes bound to reaction wells by immobilized monoclonal antibodies to conserved HLA epitopes (Schlaf et al., 2015). Donor-specific HLA antibodies are detected by enzyme labeled antihuman immunoglobulin antibodies in a colorimetric assay. Controls are incorporated for adequacy of binding of donor HLAs to the wells and nonspecific binding of the secondary antibody. The technique is reported to reduce false-positive crossmatches caused by immunotherapy treatments directed against non-HLA cell surface antigens such as anti-CD20 and anti-CD25.

In contrast to the antigen extraction procedure of Schlaf et al. (2015), a combination donor cell and target-antigen binding bead assay has been introduced to increase the HLA specificity of the crossmatch in solid organ donor crossmatches (Fig. 19B) (Chen et al., 2020). The DSA-FXM assay steps begin like a "traditional" flow cytometric crossmatch with the addition of recipient serum to isolated donor lymphocytes, incubation, and washing to remove unbound antibody. Fluorescently labeled antihuman

IgG antibody suspended in a lysis buffer is accompanied by a set of three fluorescently labeled microbeads conjugated to monoclonal antibodies directed against monomorphic epitopes on HLA Class I (one bead) or Class II proteins (two beads). The immune complexes that are formed contain donor-specific HLA antibody (if present), the fluorescently labeled anti-human IgG, and the fluorescently labeled isolation beads binding the HLA proteins from the lysed donor lymphocytes. The isolation beads are washed and analyzed by flow cytometry to detect bead alone (no antihuman IgG bound, therefore no DSA), or beads carrying recipient antibody which is specific for donor HLA. As discussed previously in this chapter, the fluorescence intensity of the bound anti-IgG is interpreted to indicate the concentration of the DSA antibody. Chen et al. stated that the three capture beads ensure coverage of the clinically relevant HLA loci (HLA-A, -B, -C, -DRB1/3/4/5, -DQA1, -DQB1, -DPA1, and -DPB1), and the antigens are presented on the beads in their natural conformations. A critical distinction of this assay is that non-HLA antibodies such as autoantibodies to structural proteins and auto- and alloantibodies to non-HLAs on lymphocytes are not detected ensuring specificity to HLA specificities in the flow crossmatch.

Note: Lymphocyte-based methods of crossmatching and antibody identification will not detect antibodies to antigens that are not expressed on lymphocytes. Although this should be obvious, lymphocytes have been the surrogate targets for crossmatching and antibody detection since human organ transplantation was introduced. Non-HLA, nonlymphocyte antigens have been shown to be associated with transplant rejection and should be considered when rejection is present and HLA antibody is not detectable (see excellent review by Zhang and Reed, 2016).

The so-called "virtual crossmatch" (or electronic crossmatch) is discussed in Chapter 5. Virtual crossmatching is dependent on the high confidence in detection and identification of HLA antibody specificity that has been enabled by immobilization of HLAs to solid-phase surfaces and highly sensitive detection of antibody binding. Serological and flow cytometric crossmatches are unable to detect which donor antigens are being detected by the potential recipient's antibodies.

References

42CFR493.1278(e). https://www.ecfr.gov/cgi-bin/text-idx?SID=1248e3189da5e5f936e55315402bc38b&node=pt.42.5.493&rgn=div5. (link confirmed 29 May 2019).

Amos, D.B., Bashir, H., Boyle, W., MacQueen, M., Tiilikainen, A., 1969. A simple micro cytotoxicity test. Transplantation 7 (3), 220–223.

Amos, D.B., Corley, R., Kostyu, D., Delmas-Marsalet, Y., Woodbury, M., 1973. The serologic structure of HL-A as indicated by crossreactivities: the inclusion of HL-A loci into major histocompatibility complex H-1. In: Dausset, J., Colombani, J. (Eds.), Histocompatibility Testing 1972. Munksgaard, Copenhagen, pp. 359–366.
Böyum, A., 1968. Isolation of mononuclear cells and granulocytes from human blood. Isolation of mononuclear cells by one centrifugation, and of granulocytes by combining centrifugation and sedimentation at 1 g. Scand. J. Clin. Lab. Invest. Suppl. 97, 77–89.
Bray, R., 2014. Virtual Crossmatch Workgroup Report to CLIAC. http://ftp.cdc.gov/pub/CLIAC_meeting_presentations/pdf/Addenda/cliac1114/8_BRAY_Virtual_Crossmatch_Workgroup_Report_Nov-2014.pdf.
Brown, J., McCloskey, D.J., Navarrete, C., 1993. HLA-DR and –DQ serotyping. In: Hui, K.M., Bidwell, J.L. (Eds.), Handbook of HLA Typing Techniques. CRC Press, Boca Raton, pp. 263–269.
Butterfield, J.H., Eisenbrey, A.B., Gleich, G.J., 1982. Membrane marker characterization of the eosinophil colony-forming cell. Br. J. Haematol. 51, 209–216.
Bux, J., Spengel, U., Mueller-Eckhardt, G., 1993. Quantitation of HLA class II antigens on B-cell lines. Eur. J. Immunogenet. 20 (3), 189–192.
Cecka, J.M., 2003. Interview with Dr Paul Terasaki. Am. J. Transplant. 3, 1047–1051.
Ceppellini, R., Dray, S., Edelman, G., 1964. Nomenclature for human immunoglobulins. Bull. World Health Organ. 30 (3), 447–450.
Chen, G., Lin, L., Tyan, D.B., 2020. DSA-FXM: accelerated donor specific flow crossmatch discriminating class I and II antibody specifically and only to donor HLA for determining true incompatibility. Transplantation 104 (4), 813–822.
Coombs, R.R.A., Mourant, A.E., Race, R.R., 1945. A new test for the detection of weak and "incomplete" Rh agglutinins. Br. J. Exp. Pathol. 26, 255–266.
Dausset, J., 1962a. The leukoagglutinins. Transfusion 2 (4), 211.
Dausset, J., 1962b. The leukoagglutinins. Transfusion 2 (4), 209–215.
Delmas-Marsalet, Y., Woodbury, M., Corley, R., Amos, D.B., 1974. Réaction de microlymphocytotoxicité avec antiglobuline anti-IgG et réactivité croisée des antigens HL-A. Nouv. Rev. Fr. Hematol. 14 (3), 409–416.
Djamali, A., Kaufman, D.B., Ellis, T.M., Zhong, W., Matas, A., Samaniego, M., 2014. Diagnosis and management of antibody-mediated rejection: current status and novel approaches. Am. J. Transplant. 14, 255–271.
Falini, B., Taylor, C.R., 1983. New developments in immunoperoxidase techniques and their application. Arch. Pathol. Lab Med. 107 (3), 105–117.
Feucht, H.E., Schneeberger, H., Hillebrand, G., Burkhardt, K., Weiss, M., Riethmüller, G., Land, W., Albert, E., 1993. Capillary deposition of C4d complement fragment and early renal graft loss. Kidney Int. 43 (6), 1333–1338.
Garovoy, M.R., Rheinschmidt, M.A., Bigos, M., Perkins, H., Colombe, B., Feduska, N., Salvatierra, O., 1983. Flow cytometry analysis: a high technology crossmatch technique facilitating transplantation. Transplant. Proc. 15 (3), 1939–1944.
Giorno, R., 1984. A comparison of two immunoperoxidase staining methods based on the avidin-biotin interaction. Diagn. Immunol. 2 (3), 161–166.
Haas, M., Sis, B., Racusen, L.C., Solez, K., Glotz, D., Colvin, R.B., Castro, M.C., David, D.S., David-Neto, E., Bagnasco, S.M., Cendales, L.C., Cornell, L.D., Demetris, A.J., Drachenberg, C.B., Farver, C.F., Farris 3rd, A.B., Gibson, I.W., Kraus, E., Liapis, H., Loupy, A., Nickeleit, V., Randhawa, P., Rodriguez, E.R., Rush, D., Smith, R.N., Tan, C.D., Wallace, W.D., Mengel, M., 2014. Banff 2013 meeting report: inclusion of C4d-negative antibody-mediated rejection and antibody-associated arterial lesions. Am. J. Transplant. 14, 272–283.
Johnson, A.H., 1976a. Antiglobulin cross-match for transplantation. In: Rose, N.R., Friedman, H. (Eds.), Manual of Clinical Immunology. American Society for Microbiology, Washington, DC, pp. 814–819.

Jones, K.H., Senft, J.A., 1985. An improved method to determine cell viability by simultaneous staining with fluorescein diacetate-propidium iodide. J. Histochem. Cytochem. 33 (1), 77–79.
Kalodimou, V., 2013. Basic Principles in Flow Cytometry. AABB Press, Bethesda, p. 1.
Kooijmans-Coutinho, M.F., Hermans, J., Schrama, E., Ringers, J., Daha, M.R., Bruijn, J.A., van der Woude, F.J., 1996. Interstitial rejection, vascular rejection, and diffuse thrombosis of renal allografts: predisposing factors, histology, immunohistochemistry, and relation to outcome. Transplantation 61 (9), 1338–1344.
Kuypers, D.R.J., Lerut, E., Evenepoel, P., Maes, B., Vanrenterghem, Y., Van Damme, B., 2003. C3d deposition in peritubular capillaries indicates a variant of acute renal allograft rejection characterized by a worse clinical outcome. Transplantation 76 (1), 102–108.
Lipshutz, G.S., McGuire, S., Zhu, Q., Ziman, A., Davis, R., Goldfinger, D., Reed, E.F., Wilkinson, A.H., Danovitch, G.M., Pham, P.-T., 2011. ABO blood type-incompatible kidney transplantation and access to organs. Arch. Surg. 146 (4), 453–458.
Maeda, M., Van Schie, R.R.A.A., Yűksel, Greenough, A., Fanger, M.W., Guyre, P.M., Lydyard, P.M., 1996. Differential expression of Fc receptors for IgG by monocytes and granulocytes from neonates and adults. Clin. Exp. Immunol. 103, 343–347.
Mallone, R., Mannering, S.I., Brooks-Worrell, B.M., Durinovec-Belló, I., Cilio, C.M., Wong, F.S., Schloot, N.C., 2010. Isolation and preservation of peripheral blood mononuclear cells for analysis of islet antigen-reactive T cell responses: position statement of the T-Cell workshop committee of the Immunology of diabetes society. Clin. Exp. Immunol. 163, 33–49.
McCloskey, D.J., Brown, J., Navarrete, C., 1993. Serological typing of HLA-A, -B, and -C antigens. In: Hui, K.M., Bidwell, J.L. (Eds.), Handbook of HLA Typing Techniques. CRC Press, Boca Raton, pp. 183–185.
Moretta, L., Ferrarini, M., Mingari, M.C., Moretta, A., Webb, S.R., 1976. Subpopulations of human T cells identified by receptors for immunoglobulins and mitogen responsiveness. J. Immunol. 117, 2171–2174. http://www.jimmunol.org/content/117/6/2171.
Perkins, W.D., Karnovsky, M.J., Unanue, E.R., 1972. An ultrastructural study of lymphocytes with surface-bound immunoglobulins. J. Exp. Med. 135, 267–276.
Ray, J.G., Hare, D.B., Pederson, P.D., Mullally, D.I., 1976. NIAID Manual of Tissue Typing Techniques. NIH, Bethesda, pp. 22–24.
Regele, H., Exner, M., Watschinger, B., Wenter, C., Wahrmann, M., Österreicher, C., Säemann, M.D., Mersich, N., Hörl, W.H., Zlabinger, G.J., Böhmig, G.A., 2001. Endothelial C4d deposition is associated with inferior kidney allograft outcome independently of cellular rejection. Nephrol. Dial. Transplant. 16, 2058–2066.
Schlaf, G., Apel, S., Wahle, A., Altermann, W.W., 2015. Solid phase-based cross-matching as solution for kidney allograft recipients pretreated with therapeutic antibodies. BioMed Res. Int. 2015, 587158. https://doi.org/10.1155/2015/587158. Published online 2015 Jan 15.
Sybesma, J.P., Kater, L., Borst-Eislers, E., de Planque, B.A., van Soelen, T., Tuit, G., 1974. HLA antigens in kidney tissue. Localization by means of an immunofluorescence technique. Transplantation 17 (6), 576–579.
Terasaki, P.I., 2007. A brief history of HLA. Immunol. Res. 38, 139–148.
Terasaki, P.I., 2012. A personal perspective: 100-year history of the humoral theory of transplantation. Transplantation 93 (8), 751–756.
Terasaki, P.I., McClelland, J.D., 1964. Microdroplet assay of human serum cytotoxins. Nature 204, 998–1000.
Vaidya, S., Cooper, T.Y., Avandsalehi, J., Barnes, T., Brooks, K., Hymel, P., Noor, M., Sellers, R., Thomas, A., Stewart, D., Daller, J., Fish, J.C., Gugliuzza, K.K., Bray, R.A., 2001. Improved flow cytometric detection of HLA antibodies using pronase: potential implications in renal transplantation. Transplantation 71 (3), 422–428.
Widmann, F.K., 1989a. An Introduction to Clinical Immunology. FA Davis, Philadelphia, pp. 93–94.

von Willebrand, E., Lautenschlager, I., Inkinen, K., Lehto, V.-P., Virtanen, I., Häyry, P., 1985. Distribution of the major histocompatibility complex antigens in human and rat kidney. Kidney Int. 27, 616–621.

Williamson, T., 1993. Basic immunology. In: Quinley, E.D. (Ed.), Immunohematology Principles and Practice. JB Lippincott, Philadelphia, pp. 55–56 (Chapter 4).

Zhang, Q., Reed, E.F., 2016. The importance of non-HLA antibodies in transplantation. Nat. Rev. Nephrol. 12 (8), 484–495.

Further reading

Dausset, J., 1962c. The leukoagglutinins. Transfusion 2 (4), 213–214.

Dausset, J., 1962d. The leukoagglutinins. Transfusion 2 (4), 213.

Johnson, A.H., 1976b. Antiglobulin cross-match for transplantation. In: Rose, N.R., Friedman, H. (Eds.), Manual of Clinical Immunology. American Society for Microbiology, Washington, DC, p. 818.

Widmann, F.K., 1989b. An Introduction to Clinical Immunology. FA Davis, Philadelphia, p. 95.

CHAPTER 5

HLA antibody detection (screening) and identification

Contents

Bullet points:

1. Solid-phase methods of antibody detection have become the standard of practice in histocompatibility testing.
2. Cell-based human leukocyte antigen (HLA) antibody identification is obtained from reactions with panels of previously typed cells from which a panel reactive antibody (PRA) percentage is calculated.
3. The improvements in crossmatch methods apply to HLA antibody detection and identification in cell-based methods.
4. Although crossmatching is a form of antibody detection, the specificity of the antibodies cannot be determined from the reaction of one person's serum with the cells from one individual.

When lymphocytes are used as the target cells for human leukocyte antigen (HLA) antibody detection and identification, the method is, essentially, performance of multiple crossmatches between the serum or sera being tested and the cells of the individuals from whom the lymphocytes were obtained. The ability to detect antibody specificity is limited by how completely the target cells are characterized by typing (e.g., when only four

HLA from Benchtop to Bedside
ISBN 978-0-12-823976-6
https://doi.org/10.1016/B978-0-12-823976-6.00005-6

HLAs were known, only four could be "reliably" detected) and by the number of typed donors available to include in the panel of target cells. Most antibody detection and identification systems divide the major histocompatibility complex (MHC) into HLA Class I (primarily T cell) and HLA Class II (primarily B cell) targets.

Since T cells normally express HLA Class I and (normally) very little HLA Class II, the ability to detect and identify antibody to HLA Class I (HLA-A, HLA-B, and HLA-C) is very robust in cell-based assays (as is the ability to *type* for the HLA Class I antigens by serological (antibody-mediated) methods). Detection of antibody to HLA Class II by serological methods is much more complicated, much less sensitive, and much less specific because of presence of HLA Class I on B lymphocytes and the relatively lower density of the HLA Class II on the target cells.

Antibody identification by microlymphocytotoxicity

Although crossmatching is a form of antibody detection, the specificity of the antibodies cannot be determined from the reaction of one person's serum with the cells from one individual. Using a "panel" of cells from multiple donors who were previously HLA "typed," the pattern of positive and negative reactions can be analyzed to determine what antibodies to HLA specificities are present in the serum. In practice, the cells from individual donors were loaded into each well of a tray (Figure 13), and the serum being tested is added to each well. As with the crossmatch, a positive control well using a pan-reactive serum and a negative control well using known negative serum are included to ensure validity of the reactions on the tray.

Selection of donors for an antibody identification cell tray is tedious. Cell donors must be selected to ensure representation of the 100 recognized HLA Class I specificities and 39 recognized HLA Class II specificities. Many laboratories would use "walking panels" of laboratory and hospital staff members who had been previously HLA typed from whom target lymphocytes could be isolated from fresh whole blood phlebotomy. A 60 well tray could have lymphocytes isolated from 58 individuals plus the two control wells (common tray formats are 30, 60, and 72 wells). Some laboratories would use preparations of frozen lymphocytes from previously typed donors and staff to construct the trays. The introduction of commercially prepared frozen cell trays was a major advance and improved the ability to compare screening results from one HLA laboratory to another (It is statistically improbable for locally prepared trays from different laboratories to have the same distribution of HLA antigens represented).

Prior to the wide-spread adoption of the commercially prepared frozen cell trays, the trays used in each HLA laboratory were intended to reflect the HLAs most likely to be found in the local community. Rare HLAs were not likely to be present on the tray unless the laboratory was fortunate to have an individual with the rare type on staff. As a result, donor cells for antibody screening trays were selected to represent the local community rather than to be comprehensive for all of the HLAs known. Cell selection for antibody identification trays was intended to not mask reactions to low frequency antigens by reactions to more common antigens. A well-designed screening tray can be used for antibody identification, and a well-designed antibody identification tray can be used for screening.

There are 139 recognized serological specificities for HLA Class I (100) and Class II (39) typing by antibody-mediated methods (HLA-A: 28; HLA-B: 62; HLA-C: 10; HLA-DR: 24; HLA-DQ: 9; HLA-DP: 6 http://hla.alleles.org/antigens/recognised_serology.html). The recognized antigens include the shared HLA class I epitopes recognized as HLA-Bw4 and HLA-Bw6 and the products of the HLA-DRB3 (DR52), HLA-DRB4 (DR53), and HLA-DRB5 (DR51) loci which are in distinct linkage groups with alleles at the HLA-DRB1 locus and were initially thought to be "public antigens" of HLA-DR.

Interpretation of antibody specificity in cell-based assays

Interpretation of antibody reactivity to screening cells is based on pattern recognition, having thoroughly typed (HLA Class I and Class II) target cells, and use of statistical methods to support the interpretation. In principle, an antibody should react with cells that have the corresponding antigen and not react with cells that lack the antigen. When a serum appears to have only one specificity, this works pretty well. It becomes much more complicated when the serum has multiple specificities. Computer programs have replaced the manual statistical analyses which were widely adopted in the 1970s but retained the chi-square (X^2) or Fisher exact test analysis of the strength of the associations of positive and negative reactions with the known HLAs of the target cells (Selwood and Hedges, 1978). Because donor lymphocytes are unlikely to be homozygous for all but for the most common HLAs, it is difficult to interpret reactions that require multiple cells to have positive reactions and alternative antigen combinations that are clearly negative. The use of software to apply the statistical tests to all of the reaction combinations for 58–70 cells both speeds analysis and improves the quality of the assignments.

Each laboratory must set minimum criteria for calling antibody screening and identification reactions positive and minimum X^2 or Fisher exact test value. In my experience as a laboratory director and inspector, most laboratories chose statistical values that gave $P \leq .10$ and individual well scores of ≥ 4. Great care was taken to ensure that all technologists reading antibody-mediated cell killing results scored consistently within one score of each other (e.g., if the expected score was 6, then acceptable results would be 4, 6, or 8). One of the many challenges for the director and supervisor was to address scoring bias when a technologist consistently gave lower (or higher) scores than the group median.

It was, and is, expected that most patients will not have antibodies to HLA. Therefore, most patients can simply be screened on a periodic basis using broadly representative reagent cells. If the screening results are positive, the patient can be tested using more cells for antibody identification. Selecting different cell donors for the screening trays and antibody identification trays allows for a larger sample size for the statistical analysis of the reactions and greater reliability in the results.

Panel reactive antibody

The number of donors' cells to which a serum would react was called the "panel reactive antibody" or PRA and was given as a percentage (%). Prior to the availability of the commercial frozen cell trays, the PRA for one serum would vary widely depending on the selection of local donors used to construct the screening trays at each HLA laboratory. Single sera sent for blind interlaboratory comparisons were occasionally found to have PRA values that would vary between zero and 100% in microcytotoxicity due to differences in technique and the panels of cells used to construct the trays (correlation coefficients ranging from 0.57 to 0.94) (Buelow et al., 1995). Buelow et al. did not find that frozen cell trays improved the correlation between laboratories even though three laboratories in their study used the same manufacturer's lot and shipment of commercially prepared frozen cell trays for the lymphocytotoxicity antibody screening assay.

Although cell-based methods for HLA antibody screening and identification have been largely replaced by solid-phase assays, immunofluorescence and microcytotoxicity methods remain in use for both research and clinical antibody testing. The few clinical laboratories that continue to use microcytotoxicity for HLA antibody screening and identification use commercially prepared and distributed frozen cell trays (unpublished data from MX-1 and MX-2 Proficiency Surveys, College of American Pathologists Histocompatibility/Identity Testing Committee, used with permission).

Some unusual reaction patterns in microcytotoxicity antibody detection were recognized as "broad" reactivity (reacting across recognized antigen groups) and "partial" reactivity (reacting with some but not all examples of a well-characterized antigen). These reaction patterns lead to the identification of shared epitopes in crossreactive antigen groups (CRAGs) (or crossreactive epitope groups: CREGs) and led to the recognition of "splits" of antigens into two or more "daughter" antigens. Additional advances in identifying the epitopes recognized by HLA antibodies came with isolation of individual HLAs and the introduction of nucleotide sequencing of the MHC.

Solid-phase antibody screening and identification

Antibodies and antigens can be linked to the surface of particles, beads, or reaction wells through covalent bonds or physical absorption to polymeric surfaces. The principles of solid-phase immunoassays have been known and widely adopted in immunology, serology, and immunochemistry since the 1960s (Catt and Tregear, 1967; Voller and Bidwell, 1976).

The technical challenges and limitations of solid-phase immunoassays are primarily in maintaining "normal" antigenicity of a target antigen bound to a surface or maintaining functionality of antibody bound to a surface. Ideally, antigenic structure should remain intact when linking target antigens to solid-phase surfaces. Likewise, antibody should be bound to the surface through the Fc portion (see footnote 1., page 35), leaving the antigen-binding regions intact. Unfortunately, like most ideals, the reality is that some antigenicity is "lost" to unintended linkage to the antigen-binding structure, and some antigenicity is lost due to binding that blocks the reactive epitopes or modifies the three-dimensional structure of the antigen molecule. An added challenge with antigens is exposure of epitopes that are normally "hidden" and not "normally" available for antibody binding. Finally, it is critical that the antigen or antibody density on the surface is optimized for the intended assay. If too much antibody or antigen is bound to the surface, there may be steric hindrance that interferes with antibody binding to antigen. Likewise, if too little antigen is bound or available for antibody binding, the available targets may become saturated, leaving unbound antibody with specificity for the target.

Solid-phase immunoassays were introduced for HLA antibody screening with tray and bead based enzyme immunoassays (EIAs) (Fig. 20A and B) which gave results comparable to cell-based antibody screening methods without the need for viable cellular material (Kao et al., 1993; Arnold et al., 2005). A disadvantage of the EIAs was the need for an EIA plate reader, common in general laboratories but unusual in HLA labs. In contrast, a flow cytometric method using purified HLAs absorbed to microbeads allowed screening for HLA antibodies on the same instrument used for performing flow crossmatches (Müller-Steinhardt et al., 2000). The flow cytometry beads are internally labeled with fixed ratios of fluorescent dyes which allow identification of bead populations to which known antigens or mixtures of antigens are absorbed/attached (Fulton et al., 1997) (Fig. 21A and B).

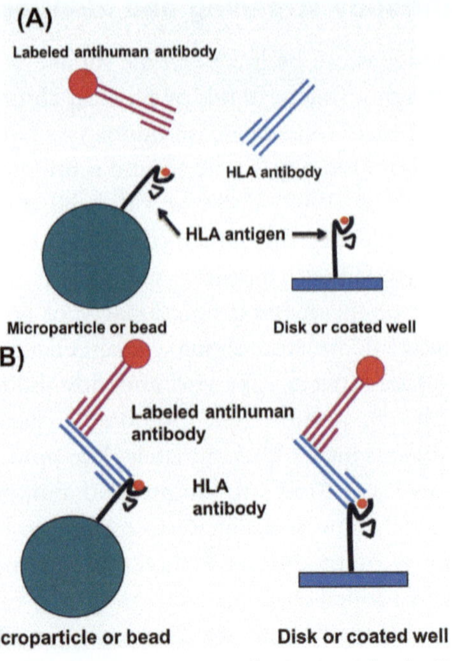

Figure 20 (A) Solid-phase detection systems for HLA antibody. Purified HLAs retain normal reactivity with antibody when beta-2 microglobulin and the bound oligopeptide are present and the antigen is attached to the surface near the carboxy terminus. Denatured protein, loss of the oligopeptide or beta-2 microglobulin, or attachment near the amino terminus of the protein may result in loss of reactivity or exposure of novel epitopes and false-positive reactions. (B) Detection of bound antibody in the solid-phase immunoassay.

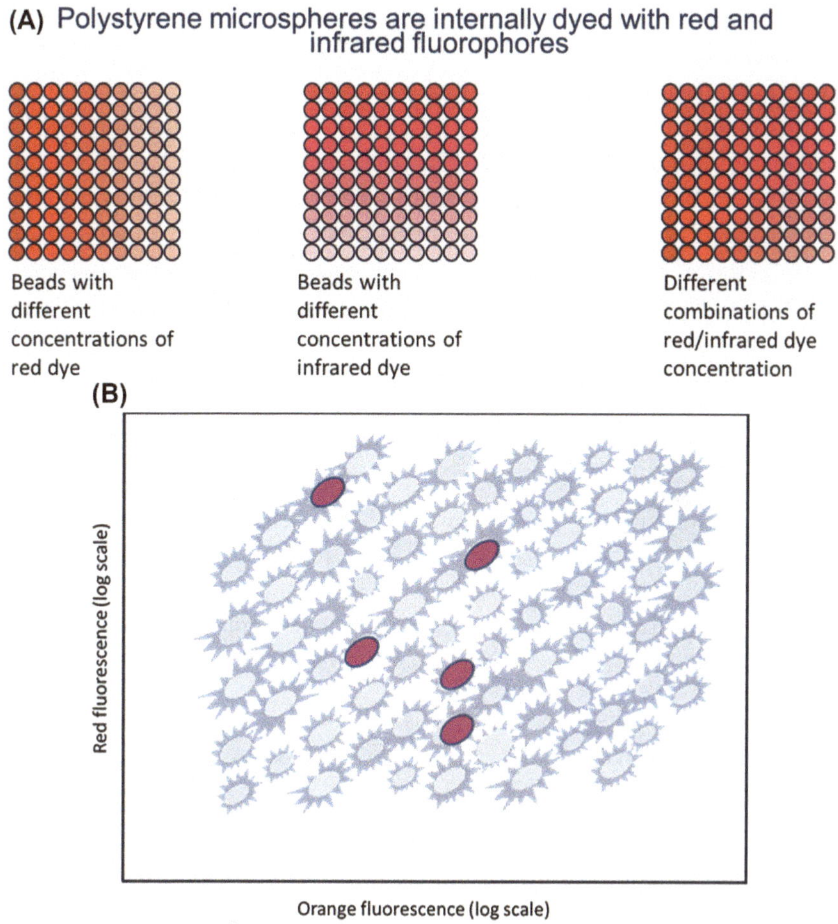

Figure 21 (A) Luminex® (Luminex Corporation, Austin, TX) dye distribution diagram. (B) Cartoon of fluorochrome classification of microspheres used in multiplex testing. Each microsphere is mapped by channel number in two dimensions by unique ratios of fluorochromes. Mapping of the beads allows identification of the specific substance(s) attached to the microbead surfaces. Note that bead aggregates are excluded in the analysis by side-scatter size analysis. *(A) Courtesy of Mariane Wolfe, Michigan State University, 2019. (B) Fulton, R.J., McDade, R.L., Smith, P.L., Kienker, L.J., Kettman, J.R., 1997. Advanced multiplexed analysis with the FlowMetrix™ system. Clin. Chem. 43(9), 1749–1756. Positive beads are indicated by purple in this cartoon.*

Solid-phase HLA antibody detection assays can be roughly "lumped" into two groups: screening and identification (similar in principle to the use of screening and identification cell trays). The original assay described by Kao et al. (1993) used HLA Class I which had been purified from pooled platelets (human platelets do not normally express HLA Class II) which were absorbed onto the surfaces of the wells of 96-well polystyrene plastic trays. The assay was functionally the equivalent to screening for HLA antibody with lymphocytes from multiple donors. The relative amount of antibody bound (as a ratio of optical density compared to a pooled serum known to be nonreactive for HLA antibody) correlated well with the PRA percentage from standard lymphocytotoxicity assays. Likewise, the bead-based flow cytometric assay ("flow beads") described by Müller-Steinhardt et al. (2000) and the bead-based EIA described by Arnold et al. (2005) were also screening assays and correlated well with results from cell-based lymphocytotoxicity assays. Flow bead assays continue to be viable for screening for HLA Class I and Class II antibodies.

Solid-phase assays for HLA antibody identification can be constructed with purified HLAs from donor cells and cell lines with known HLA phenotypes and, ultimately, from recombinant single antigens bound to individual beads (single antigen beads (SAB)). The latter assay format is available from commercial vendors for the Luminex® (Luminex Corporation, Austin, TX) multiplex fluorometric microsphere flow cytometer instruments. An FDA (Food and Drug Administration)-approved EIA method for screening blood donors for HLA antibody is available. EIA kits for HLA antibody identification are no longer available.

Commercial solid-phase HLA antibody assays are qualitative, not quantitative

Although it is possible to generate standard curves for HLA antibody tests using reagents and immunoglobulins from vendors other than the HLA antibody detection kits (e.g., 56834 SIGMA, Sigma-Aldrich, St Louis, MO), the manufacturers of the commercial solid-phase HLA antibody assays in the United States have only requested FDA approval of the assays as **qualitative** assays for antibody. Using the kits for quantitation of HLA antibody becomes a "laboratory developed test" (LDT) under Clinical Laboratory Improvement Amendments (CLIA) regulations and FDA guidance (http://www.fda.gov/downloads/medicaldevices/deviceregulationandguidance/guidancedocuments/ucm416685.pdf; http://www.cms.gov/Regulations-and-Guidance/Legislation/CLIA/Downloads/LDT-and-CLIA_FAQs.pdf). Monoclonal antibodies are also available from multiple vendors to quantitate HLA Class I (e.g., W6/32) (Barnstable et al., 1978) and HLA Class II (CR3/43) (Sunderland et al., 1981). The monoclonal antibodies can be used to quantify solid-phase antigen density and compensate for differences in HLA antibody binding intensity from well-to-well or bead-to-bead.

There are two vendors of solid-phase HLA antibody detection (screening) and identification kits in the United States (August 2018): Immucor LIFECODES, Immucor, Norcross, GA, and One Lambda LABScreen®, Thermo Fisher Scientific, Waltham, MA. Both vendors use proprietary software (e.g., HLA Fusion™, Thermo Fisher Scientific, Waltham, MA) to assist in interpretation of the antibody-binding patterns to discern specificities. The Luminex® instrument software provides the median fluorescence intensity (MFI) of each mapped and identified bead. (Note: Fluorescence intensities are generally not normally distributed. The median is usually a better descriptor of the central tendency of a distribution than the mean when a population is not normally distributed (although the mode may be preferable for a unimodal population)) (Zar, 1974). The reagent vendors provide the specificities for the mapped beads which are production lot specific and require assurance that software and mapping databases are kept up to date and current. (Note: The strict requirement for matching software, databases, and bead lots also applies to molecular typing technologies that utilize the Luminex® platform for signal detection, also available from the same HLA clinical testing vendors.)

The principles of HLA antibody identification on the Luminex® platform are discussed with Figs. 22 and 23. *This is not an endorsement of any vendor.* The MFI of each bead is displayed on a bar chart with the corresponding HLA bound to the bead. Options include vendor-defined and user-defined cutoffs for "calling" individual beads positive or negative. The manufacturer's directions for the assay are for qualitative (positive and negative) determination of the presence of HLA Class II antibody.

The left panel of Fig. 22A displays the results from a sample that is clearly negative for HLA antibody in the SAB assay. The positive control bead has an MFI of 10,896.87 and the negative control bead is 29.53. The users have set the cutoffs at MFIs of 300, 2000, 5000, and 10,000 units. The right panel shows the reaction pattern of a serum with multiple anti-HLA antibody specificities using the same user-defined cutoffs. The positive control for this sample was 13,245.92, negative control 126.61, and positive/negative ratio of 104.62. Unequivocal positives are present for HLA-A3, 25, 26, 32, and 66 and HLA-B49, 51, 52, 53, 57, 58, 59, and 63. Unequivocal negatives are present for HLA-B46, 64, 65, and 76 and HLA-Cw5, 6, 8, 12, 16, and 17. With user-defined cutoffs, it is the HLA laboratory director's responsibility to determine the significance of reactive beads above each of the different cutoffs.

In the example (Fig. 22B bottom right and left) of a positive sample, multiple specificities were identified. In the example, antibodies are identified to HLA-A*02:01, A*02:02, A*02:03, A*02:05, A*68:01, A*68:02, A*23:01, A*24:02, A*24:03, B*57:01, and B*58:01.

(A)

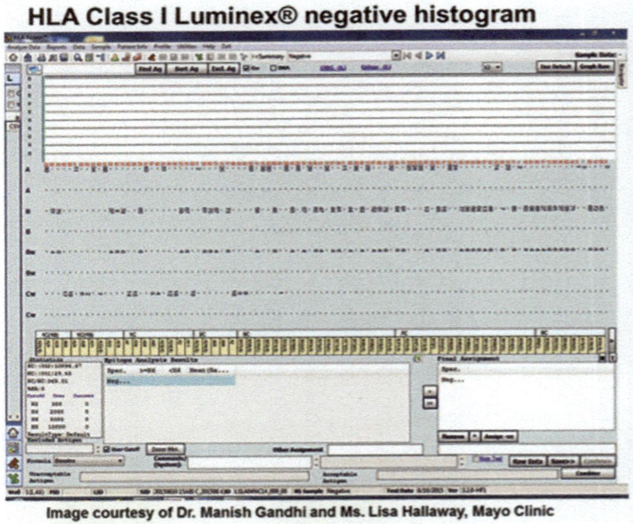

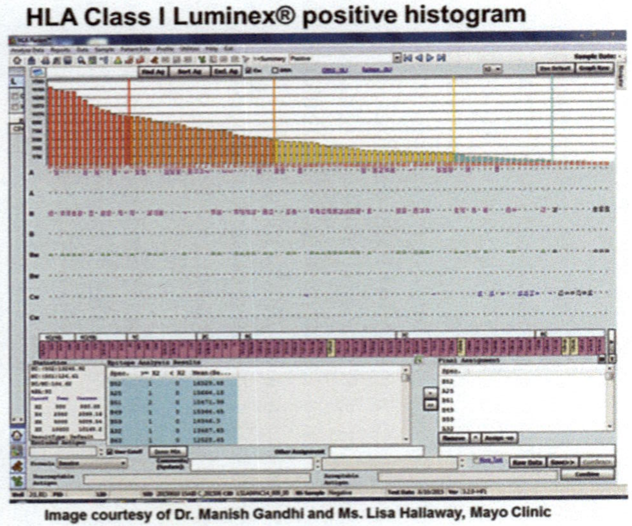

Figure 22 (A) HLA Class I single antigen bead analysis screens (HLA Fusion™). As used in these examples, cutoffs for the median fluorescence intensity (MFI) are set by the user for determination of positive, negative, and borderline reactivities. Reactive beads are assigned by the software based on either the default proprietary software using fluorescence intensity comparisons with control beads or semimanual determination by user based on raw MFI cutoffs. *(Images provided courtesy of Dr. Manish Gandhi and Ms. Lisa Hallaway, Mayo Clinic, Rochester, MN. Used with permission.)*

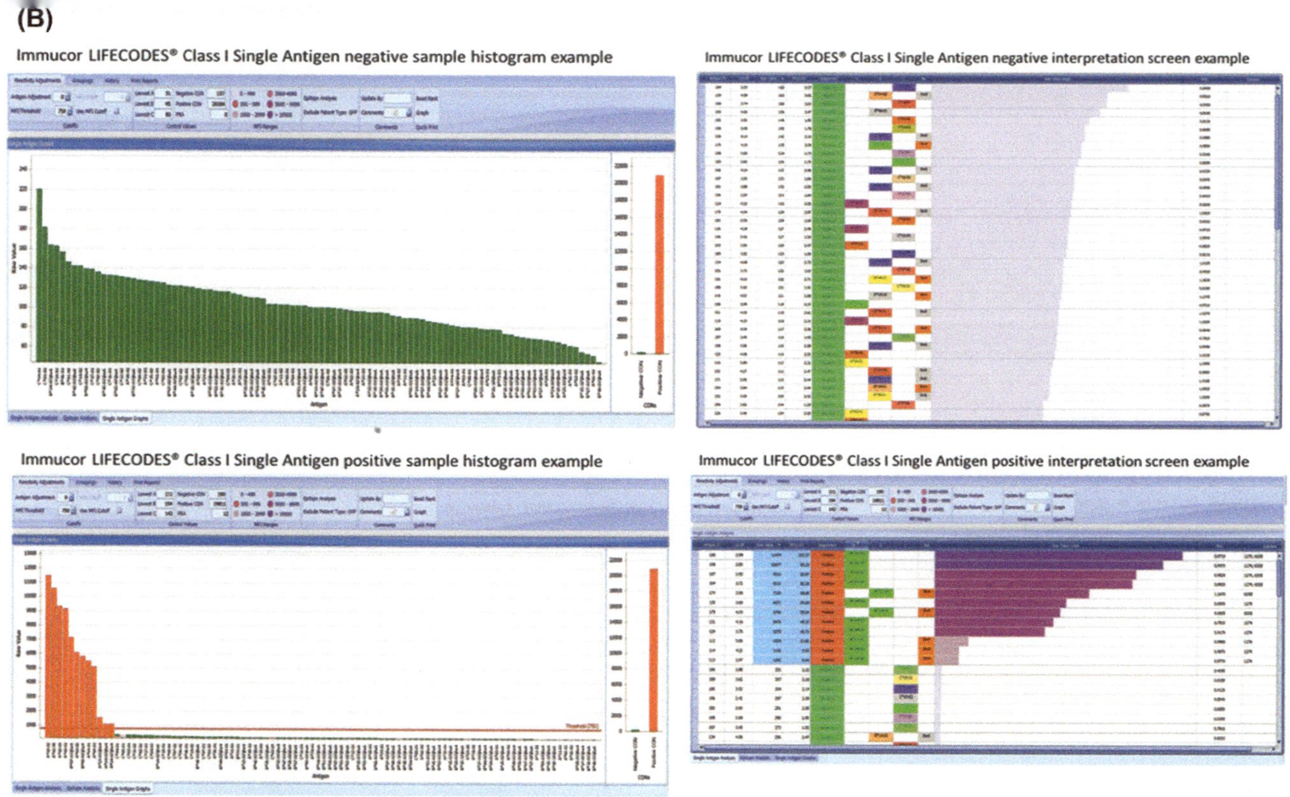

Figure 22 Cont'd (B) HLA Class I single antigen bead analysis screens (Immucor LIFECODES®). The analysis software uses manufacturer-defined cutoffs for determination of positive and negative. Note that the scale in the first, upper left (negative) histogram (high bead approximately 220 units) is much lower than the scale in the lower positive histogram (high bead approximately 11,500 units). Each histogram is followed by an image of the interpretation screen to the right. *(Images provided courtesy of Christie Otis, Immucor, Inc. Used with permission.)*

(A)

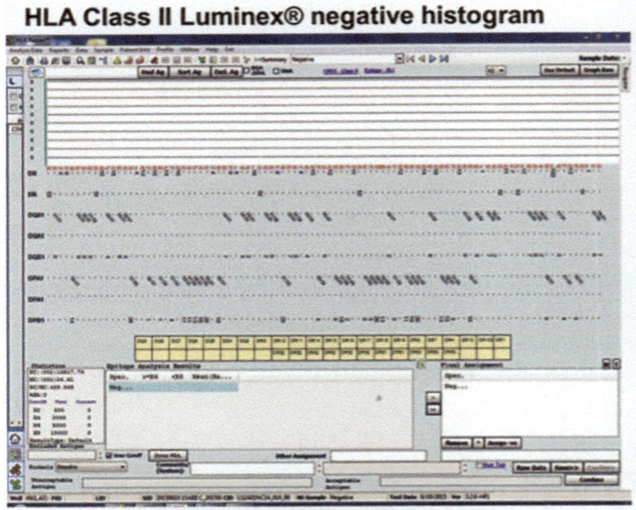

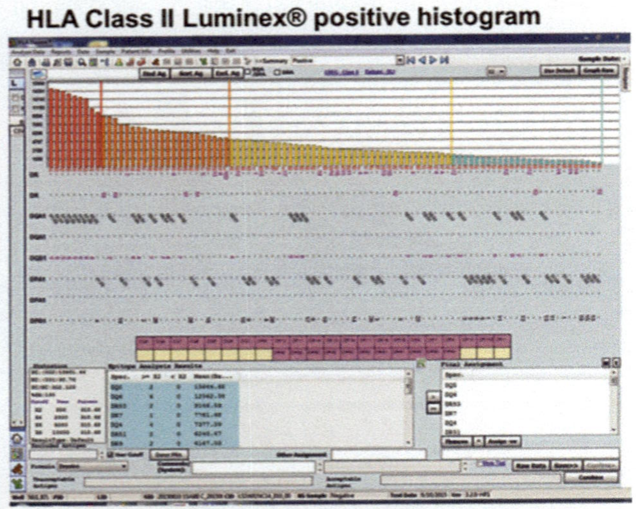

Figure 23 (A) HLA Class II single antigen bead analysis screens (HLA Fusion™). The left panel is an example of the bead pattern from a clearly negative sample. The right panel is an example of a sample with multiple antibodies to HLA Class II specificities. The user-defined cutoffs are the same as in Fig. 22A (see text). *(Images provided courtesy of Dr. Manish Gandhi and Ms. Lisa Hallaway, Mayo Clinic, Rochester, MN. Used with permission.)*

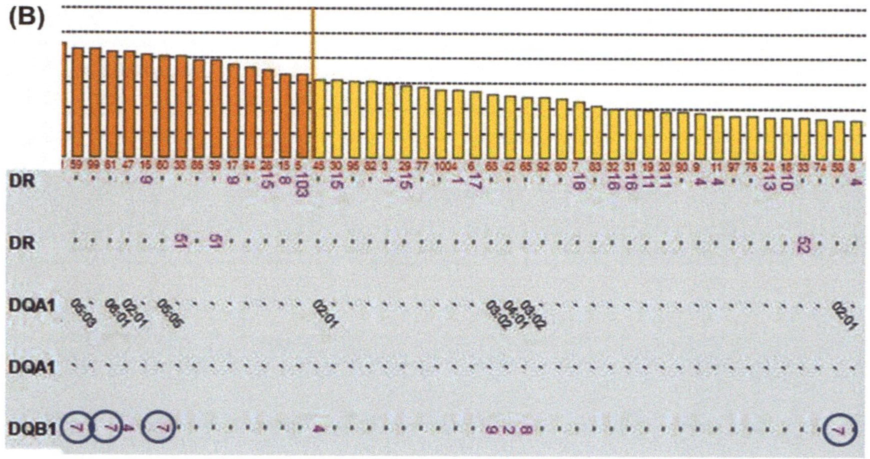

Figure 23 Cont'd (B) Close up of a portion of the right panel of Fig. 23A demonstrating HLA-DQA1/DQB1 chain effect on antibody binding. HLA-DQ7 beads are circled.

(C) Immucor LIFECODES® Class II Single Antigen positive sample histogram screen example

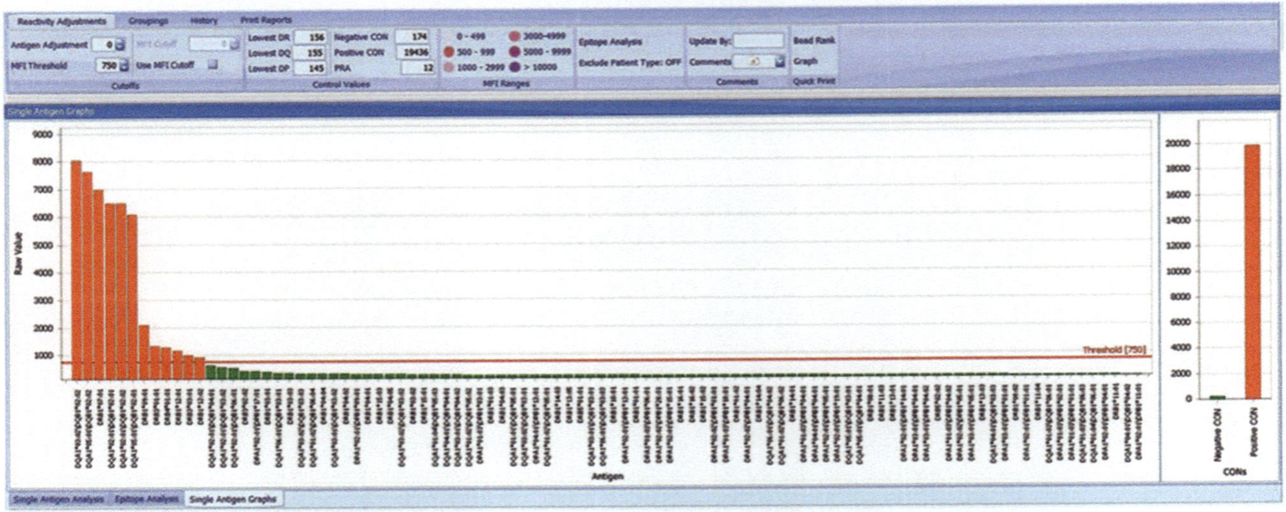

Figure 23 Cont'd (C) HLA Class I single antigen bead analysis screens (Immucor LIFECODES®). (*Negative example not shown. See Fig. 22B for an example of a HLA Class I negative reaction.*) In the example below, positive beads have raw values of just over 900 and range up to just over 8000 units. The HLA alleles of the proteins bound to the beads are listed. Antibodies are identified to DRB1*07:01, *09:01, *12:01, *12:02, DRB3*01:01, *03:01, DRB4*01:01, DQB1*02:01/DQA1*02:01, DQB1*02:01/DQA1*05:01, DQB1*02:02/DQA1*02:02, DQB1*02:02/DQA1*03:02, and DQB1*02:02/DQA1*05:01. *(Images provided courtesy of Christie Otis, Immucor, Inc. Used with permission.)*

Immucor LIFECODES® Class II Single Antigen positive interpretation screen example

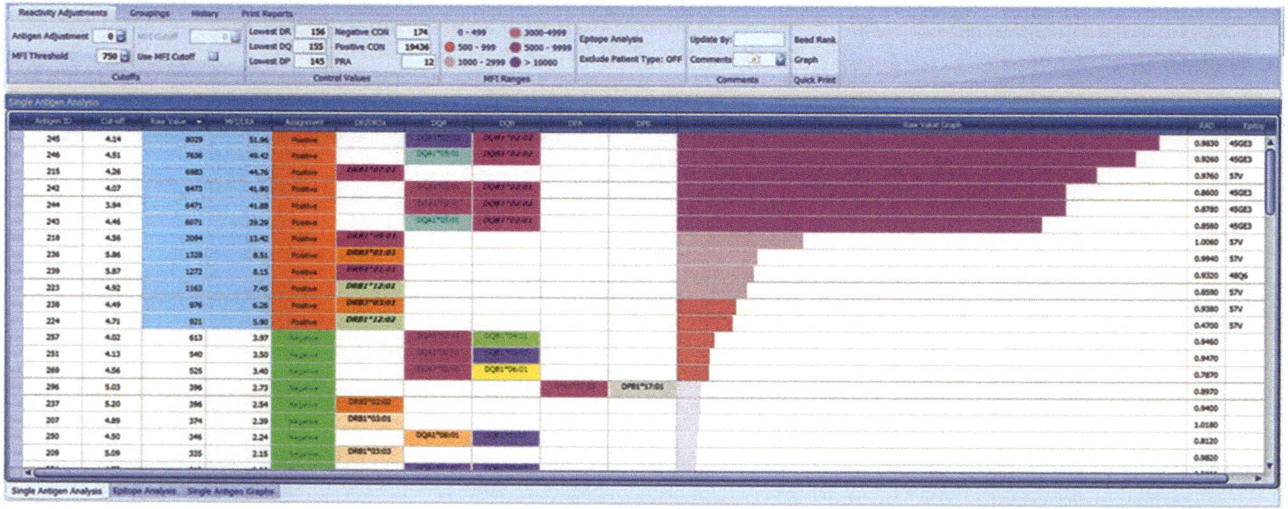

Figure 23 Cont'd (C)

The SAB assay for HLA Class II antibody (Fig. 23A) is performed and interpreted in the same manner as the SAB assay for HLA Class I antibody (Fig. 22A). The manufacturer's directions for the assay are for qualitative (positive and negative) determination of the presence of HLA Class II antibody. Options include vendor-defined and user-defined cutoffs for "calling" individual beads positive or negative and assigning HLA antibody specificities.

In the examples shown in Fig. 23A, the users have defined cutoffs of 300, 2000, 5000, and 10,000 MFI. The left panel is from a sample with no detectable HLA Class II antibody and positive control of 10,817.79, negative control of 24.61, and positive/negative ratio of 439.569. The right panel is from a sample with multiple HLA Class II antibody specificities. The positive control is 10,831.46, negative control is 30.76, and positive/negative ratio is 352.128. The sample has unequivocal antibody reactivity for HLA-DQ4, 5, and 6 although there are significantly different MFIs between beads. The lowest and highest beads for HLA-DQ4 beads differ by approximately 5000 MFI units. Because of the interaction of the HLA-DQB1 and HLA-DQA1 chains, some HLA-DQ antigen beads may be clearly negative while others are strongly positive, as in this example (Fig. 23B) for HLA-DQ7 in which the bead with the HLA-DQA1*02:01 alpha chain is nonreactive while the HLA-DQ7 beads with other HLA-DQA1 specificities are positive.

Clinical indications for HLA antibody testing

HLA antibody screening and identification are **not** routine clinical laboratory tests. The clinical indications for testing for HLA antibodies are limited to patients awaiting transplantation, posttransplant monitoring, support of patients who have become refractory to platelet transfusions, investigation of possible alloimmune thrombocytopenia, and investigation of probable transfusion-related acute lung injury (TRALI). A related but nonclinical use of HLA antibody screening is for prevention of TRALI by screening blood donors for HLA antibody (see section on HLA and Transfusion). *Requests for HLA antibody testing in other patient settings are likely to be mistakes and should be confirmed prior to testing.*

The most common indications for HLA antibody testing are related to solid organ transplantation. Although the significance of HLA antibody in the outcome of renal transplants was identified early in solid organ transplant (Patel and Terasaki, 1969), there was a slow transition from simply performing pretransplant crossmatches to the current practice of pretransplantation antibody screening of patients who are candidates for transplantation of solid organs and testing for donor-specific antibody (DSA) posttransplantation as a part of monitoring for graft rejection (Kannbhiran et al., 2015). Antibody to donor HLAs are now recognized as significant for predicting organ failure for kidney (Kannabhiran, 2015), pancreas (Mittal et al., 2014), heart (Kobashigawa et al., 2011), lung (Hachem, 2012), liver (Wozniak et al., 2015), and composite tissue (e.g., face and hand) transplants (Weissenbacher et al., 2014).

Results of pretransplant antibody identification have been included in the calculations used to assign the points (in the United States, https://optn.transplant.hrsa.gov/learn/professional-education/kidney-allocation-system/) for renal transplant candidates awaiting deceased donor organs, and to plan pre-and posttransplantation immunosuppression for all solid organ and composite-tissue transplant recipients. The decision to transplant in the presence of a preformed antibody against donor HLAs is a *clinical* decision in which the severity of the recipient's disease is weighed against the efficacy of immunosuppression and likelihood of a "better" donor. As noted in the multiple references in the previous paragraph, the presence of HLA antibody against donor antigens is predictive of accelerated, delayed, and chronic rejection.

The clinical transition to SAB assays and the ability to detect and identify HLA antibody with great sensitivity and specificity resulted in the recognition that the laboratory reported PRA was not consistent in representing the sensitization of patients to HLA. A calculated PRA (cPRA) tool was introduced by the United Network for Organ Sharing (UNOS) in 2009 (Cecka, 2010). The cPRA used the HLA antibodies detected by the individual transplant laboratory for its patients and the frequency of the identified HLAs in allocated kidney donors.

The "gold standard" against which all HLA antibody testing is measured continues to be the microlymphocytotoxicity crossmatch. In the United States, Federal Regulations require "use (of) a technique(s) that detects HLA-specific antibody with a specificity equivalent or superior to that of the basic complement-dependent microlymphocytotoxicity assay." (42 CFR 493.1278(d) (1)). The antihuman globulin (AHG) augmented microlymphocytotoxicity assay has become the de facto base measure because it, too, is enshrined in the US Federal Regulations which require "use (of) a technique(s) documented to have increased sensitivity in comparison with the basic complement-dependent microlymphocytotoxicity assay." (42 CFR 493.1278(e) (1)). As a result, whether appropriate or not, validation and clinical utility of the markedly more sensitive solid-phase and flow cytometric methods of HLA antibody detection are both measured against and clinically correlated with the cell-based microlymphocytotoxicity methods (Bray and Gebel, 2009).

Quantitation of "how much" antibody is significant remains a challenge, particularly since the US FDA has not approved solid-phase test systems for quantitation of HLA antibody (see following paragraph). High MFI antibody to donor HLAs has been demonstrated to be more predictive of poor graft outcomes in both solid organ and stem cell transplantation when compared to simple presence or absence of antibody (Ruggeri et al., 2013; Lefaucheur et al., 2010; O'Leary et al., 2011; Tambur et al., 2015). (Note: "median" and "mean" are often used interchangeably when discussing or reporting MFI. As noted previously, the commonly used instrumentation reports median values for the fluorescence which is a better descriptor for the intensity because the intensity is not normally distributed around the peak.)

When performed according to the manufacturers' directions, the commercial single antigen HLA antibody assays are powerful methods for detecting and tracking HLA antibody *specificities* in patients but are not informative about changes in the concentration of antibody to HLAs. When HLA density from bead to bead is considered (Hilton and Parham, 2013), tracking sample to sample changes in the MFI for individual patients and changes in relative concentration of HLA antibody can be determined and followed. For laboratories in the United States, using changes in MFI to track changes in antibody concentration (strength or "titer"[1]) instead of the qualitative presence or absence of antibody requires that the test has to be validated as a LDT in each laboratory in which it is used (this may also apply to non-USA laboratories that are accredited by ASHI or CAP).

An international consensus conference on HLA antibody testing, convened by The Transplantation Society in 2012 effectively summarized the status of clinical testing for HLA antibodies and compared and contrasted cell-based and solid-phase assays (Tait et al., 2013). The authors state that SAB arrays "are the most sensitive and specific, providing the highest degree of HLA antibody resolution, and are particularly useful in

[1] "Titer" is frequently misused in clinical science to refer to concentration. A titer is a dilution of a sample at which a defined reactivity is no longer detected. For example, a red cell antibody titer may be the dilution at which reactivity, defined as agglutination, is no longer detected. Titer is always method-dependent, and comparisons of titers must be done with the same method and targets.

the accurate identification of antibodies in highly sensitized patients." Although the manufacturers' directions for the "multiplexed multianalyte bead arrays" instruct the use of the assays as qualitative, the authors of the consensus report noted that the tests are semiquantitative because antibody binding is "expressed as mean fluorescence intensity (MFI) of the reporter signal."

An unanticipated result of the SAB technology has been the identification of allele-specific HLA antibodies within antigen groups (e.g., an A*24:02 antibody in a patient who is HLA-A*24:03 positive) (Tait, 2016). Identification of allele-specific antibodies is similar to the effect of earlier improvements in the resolution of antibody detection which first allowed for locus-related antigen/antibody identification and incrementally improved to detection of antigen "splits" which were, in time, recognized as unique antigens. Users of microbead fluorochromatography antibody detection/identification systems will see both panels expand and resolution continue to improve. There are strong arguments for defining HLA antibody specificity in terms of epitopes (unique or private vs. public or shared) rather than antigens, antigen groups, and crossreactivity (El-Awar et al., 2017).

Single antigen bead result interpretation in clinical practice

Interpretation of SAB antibody detection results is complicated by the fact (as noted previously) that the US FDA has only approved the commercial assays as qualitative, not quantitative. The reader is reminded that following the manufacturers' directions in the use of the assays and the accompanying manufacturer-provided software is in compliance with both regulatory and accreditation requirements. If the laboratory director, technical supervisor, clinical consultant, and transplant service customers find this to be adequate, no further discussion is needed. In clinical practice, physicians and clinical laboratorians find representations of the strength of an HLA antibody relevant to patient care and will seek some parameter that indicates strength or concentration.

In clinical practice, HLA antibody interpretation is challenging because correlation between *detection* of antibody and the *clinical significance* of detected antibody is not always clear. As noted previously, a positive HLA crossmatch (lymphocytotoxicity or flow) is predictive of increased *likelihood* (risk) of both acute and chronic rejection in solid organ transplantation.

Increased likelihood does not mean that an individual patient with evidence of antibody to donor antigens will acutely reject the organ or develop chronic rejection. Anecdotally, I have recommended against transplant when crossmatches were reproducibly positive and a DSA was detected, the decision was made by the transplant team to proceed with the transplant and the recipient had a very good outcome. I have also seen patients with repeatedly negative HLA antibody screens and repeatably negative crossmatches develop very strong, biopsy-proven antibody-mediated rejection. These are, however, the exceptions to the rule.

Broadly, transplantation HLA antibody testing falls into three clinical groups: pretransplant antibody screening and identification, crossmatch, and posttransplantation monitoring for DSA. Crossmatching has been discussed previously, and problematic crossmatches are primarily those that are unexpected because of good HLA matching between the donor and recipient or the lack of a history of sensitizing events in the proposed recipient (e.g., a young male who has never been transfused and had not received a previous transplant). Similarly, HLA antibodies are expected to reflect the patient's history of sensitizing events. There is no surprise when a multiparous woman is found to have HLA antibodies. There is surprise and concern when a male with no history of sensitizing events is found to have HLA antibodies.

Two interlaboratory studies, using SAB kits, reagents, and procedures from one or the other of the two SAB HLA antibody vendors, demonstrated very high concordance for antibody detection, and MFI results when the same kit lot and uniform procedures were used (Oh et al., 2015; Gandhi et al., 2013). Both reports also noted that the greatest variability in MFI for individual samples and beads was found when the MFI was less than 1000. Although low MFI samples (e.g., MFI<500) are less likely to result in a positive microcytotoxicity or flow cytometric crossmatch, samples with multiple low MFI antibodies may result in a positive crossmatch (Sullivan et al., 2017). As stated by Sullivan et al. (2017, page 473), "The 'appropriate' threshold should be a programmatic decision involving the HLA laboratory and members of the transplant program."

My practice, based on clinical experience as an HLA laboratory director and my experiences as an inspector and member of the CAP proficiency testing oversight committee, was to set MFI cutoffs for calling antibody positivity based on both laboratory and clinical data. Solid-phase antibody detection results were correlated with historical microlymphocytotoxicity results and crossmatch results (primarily with flow cytometric crossmatch results). For deceased donor crossmatches, we routinely investigated any unexpected positive or negative crossmatch results to determine if cutoffs (MFI thresholds) were set too low or too high or if HLA specificities were missed on the commercial beads. Interlaboratory exchanges were used between HLA laboratories in Michigan to determine correlation between deceased donor crossmatch positivity rates and cutoffs for calling antibody positivity in the SAB assays at the participating transplant hospitals (unpublished quality assurance data). (Note: In Michigan, all deceased donor HLA typing and crossmatching is performed by the Organ Procurement Organization, Gift of Life Michigan, Ann Arbor, MI.) The SAB MFI cutoffs used in the laboratories I directed were reviewed at least annually and formalized in the written agreements with the transplant services. Perhaps not surprisingly, the cutoffs changed over time because of bead lots and changes in reagents, for example.

MFI as a quantitative result

A clinical advantage of using MFI in HLA antibody determinations is the potential for trending the relative strength of the antibody specificities. This is consistent with most of clinical laboratory science and clinical practice because physicians routinely make diagnostic and treatment decisions based on numerical quantitative and semiquantitative results. As noted previously, the commercial HLA antibody detection and identification assays are approved as qualitative, not quantitative. The vendors do not provide standards or reagents for quantifying the strength of the detected antibody.

The temptation to use the results of the SAB assays as quantitative results should be tempered with understanding of the limitations, clearly deliminated in the review by Sullivan et al. (2017), cited above. The most serious limitations include variable density of antigen bound to the reagent beads, lot to lot differences, differences between the beads of the two vendors, and significant variation between laboratories and, within some laboratories, between assays. As noted by Oh et al. (2015) and Gandhi et al. (2013), some intralaboratory variability can be controlled with strict adherence to procedures, but in most labs the testing remains largely manual and procedurally complicated. Two additional limitations of the bead assays may result in underestimating the strength (concentration) of antibody due to saturation of the available antibody binding sites on a bead when the concentration of antibody is very high and underestimating the strength of an antibody when there are interfering substances such as complement or non-IgG antibodies competing for binding sites.

Commonly used methods to reduce the risk of underestimating the concentration of HLA antibodies in the SAB assays include testing diluted samples and use of EDTA plasma samples (prevents complement activation and binding).

There is a clinical need to determine if an antibody specificity is getting stronger or is responding to clinical therapy (ideally, decreasing in strength (see previous footnote on "titer")). Because of lot to lot variation in antigen density on the beads and intraassay variability, a previous sample from the patient can be tested simultaneously with the current sample to compare the MFI of the two sample dates. Less robust methods include trending the calculated assay sample versus control ratios for the specific antigen beads for the antibodies of interest. The later, indirect method, is commonly used because of the cost of repeat testing of samples.

Conclusion: If HLA antibody identification is determined by laboratory-assigned cutoffs, the cutoffs (thresholds) must be locally validated and used with concurrence of the transplant program physicians.

Virtual crossmatch

The original purpose of crossmatching was to prevent hyperacute rejection of solid organ transplants. That purpose still exists. When HLA antibody testing was performed serologically, there was substantial risk that significant antibody may be undetected due to the limitations in cell panels and the sensitivity of the testing method. Current flow cytometric and bead/particle based assays have nearly eliminated missing clinically significant antibody, particularly when patients are tested on multiple occasions.

A "virtual crossmatch" is a clinical practice in which past test results are used to predict the likelihood of incompatibility, compatibility, or acceptable incompatibility between a patient's serum or plasma and donor cells or organs. Virtual crossmatching is similar, in concept, to the FDA approved practice of an electronic or computer crossmatch used in transfusion medicine. An electronic crossmatch requires repeatedly negative tests for red cell antibodies, recipient ABO/Rh typing, a negative history for red cell antibodies, and an FDA licensed blood bank software system that prevents release of components that would be expected to be incompatible (https://www.fda.gov/downloads/BiologicsBloodVaccines/GuidanceCompliance RegulatoryInformation/Guidances/Blood/UCM252894.pdf). The electronic (computer) crossmatch has been used effectively in Transfusion Medicine and Blood Banking for more than 2 decades (Butch et al., 1994).

Transplant programs with which I have been affiliated have used virtual crossmatches to select "acceptable" mismatches in donor-recipient pairs for heart, lung, and intestine transplants, and to select potential recipients of deceased donor kidneys *without* a pretransplant flow cytometric crossmatch. Virtual crossmatching reduces cold ischemia time, particularly for organs from remote donors, and speeds determination of acceptance when lymphocytes for crossmatching are not accessible in a timely manner (Sullivan et al., 2018; Jackson, 2014; Cheng et al., 2017). The UNOS deceased donor matching program is, in effect, a virtual crossmatch program in which potential recipients are excluded based on their transplant program reporting of HLA antibody specificities ("unacceptable antigens"). For both deceased donor and living donor renal transplant pairings, there has been a reduction in the incidence of unexpectedly positive crossmatches as a result of the sensitivity and quality of HLA antibody testing using solid-phase antibody detection, particularly the flow cytometric microbead techniques (Sullivan et al., 2018; Cecka, 2011).

The US Department of Health and Human Services Clinical Laboratory Improvement Advisory Committee (CLIAC) convened a Virtual Crossmatch Workgroup, 3–4 September 2014 (Bray, 2014), to advise CLIAC of the laboratory implementation and clinical utility of a virtual crossmatch in lieu of a "physical" crossmatch. The working group presented a number of recommendations for virtual crossmatching for solid organ transplant, including: confirmation of recipient antibody specificity for the VXM by more than one method; use of multiple solid-phase testing platforms or surrogate physical crossmatches; reconfirmation of patients' HLA antibody specificity on an ongoing basis; time frames for defining a "recent" sample should be based on patient sensitization history and transplant center practices; a serum sample must be collected and stored on day of transplant; and posttransplant DSA assessments should be performed.

The CLIAC (5–6 November 2014) recommendations after presentation of the working group discussions and recommendations were: CMS should explore: "Regulatory changes or guidance(s) that would allow virtual crossmatching to replace physical crossmatching as a pre-requisite for organ transplant; and, appropriate criteria and decision algorithms, based on the Virtual Crossmatch Workgroup input provided to CLIAC, under which virtual crossmatching would be an appropriate substitute for physical crossmatching. The determination of appropriate criteria and decision algorithms should involve a process that includes an open comment period" (CLIAC Meeting Report, 2014).

As of July 2020, virtual crossmatching for solid organ transplantation has not been recognized by the USFDA or the US Department of Health and Human Services (Centers for Medicare and Medicaid Services). Recognition and/or approval are required for reimbursement purposes for laboratory expenses and for compliance issues with CLIA. Virtual crossmatching has been widely adopted, clinically, for solid organ transplantation. Virtual crossmatches are very useful for selecting acceptable donors for highly sensitized patients, particularly for life-saving transplants like heart, lung, intestine, and liver for which donors are limited and time is critical (Cheng et al., 2017). While awaiting federal recognition or approval of virtual crossmatching, transplant clinicians, laboratorians, HLA professional, and accrediting agencies, in the United States and internationally, are incorporating virtual crossmatching into clinical practice.

Management of antibody-mediated transplant rejection

Treatment of transplant organ rejection is most commonly based on local transplant program protocols, and recommendations are limited due to the difficulty of performing randomized control studies. The pathology of solid organ transplant graft rejection has been extensively studied and, as previously discussed, characterized in consensus documents published as Banff criteria or updates (e.g., Haas et al., 2014). Treatment of T lymphocyte-mediated rejection primarily focuses on modification of the patient's current immunosuppression. In contrast, treatment of antibody-mediated rejection usually is focused on reducing the plasma concentration of preformed or de novo antibody with specificity against the donor antigens (DSA), interfering with the production of the DSA, or interfering with the functional activity of the DSA (e.g., complement activation). A recent consensus document from The Transplantation Society provides a comprehensive summary of the available clinical methods to manage AMR in renal transplant recipients (Schinstock et al., 2019).

HLA in Transfusion Medicine and Blood Banking

TRALI

A unique indication for HLA antibody screening and identification is prevention of TRALI (transfusion-related acute lung injury). TRALI is a life-threatening reaction between HLA or human neutrophil antibodies (HNA) and, usually, antigens expressed in the capillary endothelium of the recipient's lung. The symptoms usually include noncardiogenic pulmonary edema, decreased oxygenation of the blood, and respiratory failure clinically similar to acute respiratory distress syndrome (ARDS). Blood donors are screened for HLA antibody to prevent transfusion of plasma-containing blood products containing measurable amounts of HLA antibody (Williams et al., 2012). As noted previously, only one assay has US FDA approval for screening blood donors for HLA antibodies. When TRALI is suspected, blood donors may be tested to detect antibody to HLAs present in the recipient. Additional testing postevent may include screening for antibodies to neutrophil-specific antigens (HNA) (Rodey, 2003b). If only plasma is transfused, the donor should be tested for HLA and HNA antibodies and the recipient should be tested to determine if the target antigen(s) is present. If plasma-containing cellular components were transfused, both donor and recipient HLA typing and antibody status should be evaluated. HNA typing is appropriate if HLA antibody is excluded. Screening of plasma donors for HLA antibodies has substantially reduced the incidence of TRALI.

Prevention of TRALI is the goal. Treatment of TRALI is primarily supportive (Cho and Sharma, 2018).

Platelet transfusion support

In addition to the blood group antigens ABH, Lewis, I, i, and P, human platelets express HLAs Class I and the family of platelet-specific antigens (Human Platelet-specific Antigens, HPA) (Klein and Anstee, 2005). Patients may become alloimmunized to HLAs and HPAs after transfusions, transplantation, or pregnancy and fail to respond to platelet transfusions from donors with the antigen(s) to which the patient has antibody(ies). Maternal antibody to platelet antigens expressed by the fetus may result in neonatal alloimmune thrombocytopenia and risk of life-threatening hemorrhage in the newborn. A role for the HLA laboratory is to determine the presence or absence of HLA Class I antibodies in those patients who do not respond to platelet transfusions (refractoriness) and mothers of infants with thrombocytopenia. If transfusion of platelets is indicated, donors are selected who are negative for the antigens for which antibody has been detected (Rodey, 2003a; Kickler, 2003). Clinical management of patients who are refractory to platelet transfusions can be found in Transfusion Medicine textbooks and the AABB *Technical Manual*, for example (Kaufman and Shehata, 2017). Strategies for management include solid-phase crossmatching of donor platelets with the recipient's serum, selecting platelet donors who lack the HLAs Class I to which the recipient has antibody, and HLA-matching of platelet donors to the recipient.

References

Arnold, M.-L., Pei, R., Spriewald, B., Wassmuth, R., 2005. Anti-HLA class II antibodies in kidney retransplant patients. Tissue Antigens 65, 370–378.

Barnstable, C.J., Bodmer, W.F., Brown, G., Galfre, G., Milstein, C., Williams, A.F., Ziegler, A., 1978. Production of monoclonal antibodies to group A erythrocytes, HLA and other human surface antigens – new tools for genetic analysis. Cell 14 (1), 9–20.

Bray, R., 2014. Virtual Crossmatch Workgroup Report to CLIAC. http://ftp.cdc.gov/pub/CLIAC_meeting_presentations/pdf/Addenda/cliac1114/8_BRAY_Virtual_Crossmatch_Workgroup_Report_Nov-2014.pdf.

Bray, R.A., Gebel, H.M., 2009. Strategies for human leukocyte antigen antibody detection. Curr. Opin. Organ Transplant. 14, 392–397.

Buelow, R., Mercier, I., Glanville, L., Regan, J., Ellingson, L., Janda, G., Claas, F., Colombe, B., Gelder, F., Grosse-Wilde, H., Orosz, C., Westhoff, U., Voegeler, U., Monteiro, F., Pouletty, P., 1995. Detection of panel-reactive anti-HLA class I antibodies by enzyme-linked immunosorbent assay or lymphocytotoxicity. Results of a blinded, controlled multicenter study. Hum. Immunol. 44 (1), 1–11.

Butch, S.H., Judd, W.J., Steiner, E.A., Stoe, M., Oberman, H.A., 1994. Electronic verification of donor recipient compatibility: the computer crossmatch. Transfusion 34, 105–109.

Catt, K., Tregear, G.W., 1967. Solid-phase radioimmunoassay in antibody-coated tubes. Science 158, 1570–1572.

Cecka, J.M., 2010. Calculated PRA (CPRA): the new measure of sensitization for transplant candidates. Am. J. Transplant. 10, 26–29.

Cecka, J.M., Kucheryavaya, A.Y., Reinsmoen, N.L., 2011. Calculated PRA: initial results show benefits for sensitized patients and a reduction in positive crossmatches. Am. J. Transplant. 11, 719–724.

Cheng, E.Y., DuBray, B.J., Farmer, D.G., 2017. The impact of antibodies and virtual crossmatching on intestinal transplant outcomes. Curr. Opin. Organ Transplant. 22, 149–154.

Cho, M.S., Sharma, S., January 2018. Transfusion, acute lung injury. In: StatPearls [Internet]. Treasure Island (FL). StatPearls Publishing [Updated 2018 Jun 14]. Available from: https://www.ncbi.nlm.nih.gov/books/NBK507846/.

CLIAC Meeting Report, 2014. https://www.cdc.gov/maso/facm/pdfs/CLIAC/2014110506_CLIAC.pdf.

El-Awar, N., Jucaud, V., Nguyen, A., 2017. HLA epitopes: the targets of monoclonal and alloantibodies defined. J. Immunol. Res. 2017, 3406230. https://doi.org/10.1155/2017/3406230. Epub 2017 May 24.

Engvall, E., Perlmann, P., 1972. Enzyme-linked immunosorbent assay, Elisa: III. Quantitation of specific antibodies by enzyme-labeled anti-immunoglobulin in antigen-coated tubes. J. Immunol. 109, 129–135.

Fulton, R.J., McDade, R.L., Smith, P.L., Kienker, L.J., Kettman, J.R., 1997. Advanced multiplexed analysis with the FlowMetrix™ system. Clin. Chem. 43 (9), 1749–1756.

Gandhi, M.J., DeGoey, S., Falbo, D., Jenkins, S., Stubbs, J.R., Noreen, H., Lorentzen, D.F., Lee, J.H., Stegall, M., 2013. Inter and intra laboratory concordance of HLA antibody results obtained by single antigen bead assay. Hum. Immunol. 74, 310–317.

Haas, M., Sis, B., Racusen, L.C., Solez, K., Glotz, D., Colvin, R.B., Castro, M.C., David, D.S., David-Neto, E., Bagnasco, S.M., Cendales, L.C., Cornell, L.D., Demetris, A.J., Drachenberg, C.B., Farver, C.F., Farris 3rd, A.B., Gibson, I.W., Kraus, E., Liapis, H., Loupy, A., Nickeleit, V., Randhawa, P., Rodriguez, E.R., Rush, D., Smith, R.N., Tan, C.D., Wallace, W.D., Mengel, M., 2014. Banff 2013 meeting report: inclusion of C4d-negative antibody-mediated rejection and antibody-associated arterial lesions. Am. J. Transplant. 14, 272–283.

Hachem, R., 2012. Antibody-mediated lung transplant rejection. Curr. Respir. Care Rep. 1 (3), 157–161.

Hilton, H.G., Parham, P., 2013. Direct binding to antigen-coated beads refines the specificity and cross-reactivity of four monoclonal antibodies that recognize polymorphic epitopes of HLA class I molecules. Tissue Antigens 81 (4), 212–220.

http://hla.alleles.org/antigens/recognised_serology.html.

Jackson, A.M., 2014. The virtual crossmatch: an essential tool for transplanting sensitized patients. In: Everly, M.J., Terasaki, P.I. (Eds.), Clinical Transplants 2014. Terasaki Foundation Laboratory, Los Angeles (Chapter 16).

Kannabhiran, D., Lee, J., Schwartz, J.E., Friedlander, R., Aull, M., Muthukumar, T., Campbell, S., Epstein, D., Seshan, S.V., Kapur, S., Sharma, V.K., Suthanthiran, M., Dadhania, D., 2015. Characteristics of circulating donor human leukocyte antigen-specific immunoglobulin G antibodies predictive of acute antibody-mediated rejection and kidney allograft failure. Transplantation 99 (6), 1156–1164.

Kao, K.-J., Scornik, J.C., Small, S.J., 1993. Enzyme-linked immunoassay for anti-HLA antibodies – an alternative to panel studies by lymphocytotoxicity. Transplantation 55 (1), 192–196.

Kaufman, R.M., Shehata, N., 2017. Hemotherapy decisions and their outcomes. In: Fung, M.K., Eder, A.F., Spitalnik, S.L., Westhoff, C.M. (Eds.), Technical Manual, 19th Edition. AABB, Bethesda, MD, pp. 418–420 (Chapter 19).

Kickler, T.S., 2003. Transfusion of the platelet-refractory patient. In: Silberstein, L.E., Ness, P.M., Anderson, K.C. (Eds.), Blood Banking and Transfusion Medicine: Basic Principles & Practice. Hillyer CD. Churchill Livingstone, Philadelphia, PA, pp. 370–374.

Klein, H.G., Anstee, D.J. (Eds.), 2005. Mollison's Blood Transfusion in Clinical Medicine, eleventh ed. Blackwell Publishing, Malden, MA, pp. 570–573 (Chapter 13).

Kobashigawa, J., Crespo-Leiro, M.G., Ensminger, S.M., Reichenspurner, H., Angelini, A., Berry, G., Burke, M., Czer, L., Hiemann, N., Kfoury, A.G., Mancini, D., Mohacsi, P., Patel, J., Pereira, N., Platt, J.L., Reed, E.F., Reinsmoen, N., Rodriguez, E.R., Rose, M.L., Russell, S.D., Starling, R., SuciuFoca, N., Tallaj, J., Taylor, D.O., Van Bakel, A., West, L., Zeevi, A., Zuckermann, A., 2011. Report from a consensus conference on antibody-mediated rejection in heart transplantation. J. Heart Lung Transplant. 30 (3), 252–269.

Lefaucheur, C., Loupy, A., Hill, G.S., Andrade, J., Nochy, D., Antoine, C., Gautreau, C., Charron, D., Glotz, D., Suberbielle-Boissel, C., 2010. Preexisting donor-specific HLA antibodies predict outcome in kidney transplantation. J. Am. Soc. Nephrol. 21 (8), 1398–1406.

Legislation. http://www.cms.gov/Regulations-and-Guidance/Legislation/CLIA/Downloads/LDT-and-CLIA_FAQs.pdf.

Medicaldevices. http://www.fda.gov/downloads/medicaldevices/deviceregulationandguidance/guidancedocuments/ucm416685.pdf.

Mittal, S., Page, S.L., Friend, P.J., Sharples, E.J., Fuggle, S.V., 2014. De novo donor-specific HLA antibodies: biomarkers of pancreas transplant failure. Am. J. Transplant. 14, 1664–1671.

Müller-Steinhardt, M., Fricke, L., Kirchner, H., Hoyer, J., Klüter, H., 2000. Monitoring of anti-HLA class I and II antibodies by flow cytometry in patients after first cadaveric kidney transplantation. Clin. Transplant. 14, 85–89.

Oh, E.-J., Park, H., Park, K.U., Kang, E.-S., Kim, H.-S., 2015. Interlaboratory comparison of the results of lifecodes LSA Class I and Class II single antigen kits for human leukocyte antigen antibody detection. Ann. Lab. Med. 35 (3), 321–328.

O'Leary, J.G., Kaneku, H., Susskind, B.M., Jennings, L.W., Neri, M.A., Davis, G.L., Klintmalm, G.B., Terasaki, P.I., 2011. High mean fluorescence intensity donor-specific anti-HLA antibodies associated with chronic rejection postliver transplant. Am. J. Transplant. 11 (9), 1868–1876.

Patel, R., Terasaki, P.I., 1969. Significance of the positive crossmatch test in kidney transplantation. N. Engl. J. Med. 280, 735–739.

Rodey, G.E., 2003a. HLA and granulocyte antigens. In: Silberstein, L.E., Ness, P.M., Anderson, K.C. (Eds.), Blood Banking and Transfusion Medicine: Basic Principles & Practice. Hillyer CD. Churchill Livingstone, Philadelphia, PA, pp. 88–90.

Rodey, G.E., 2003b. HLA and granulocyte antigens. In: Silberstein, L.E., Ness, P.M., Anderson, K.C. (Eds.), Blood Banking and Transfusion Medicine: Basic Principles & Practice. Hillyer CD. Churchill Livingstone, Philadelphia, PA, pp. 90–91.

Ruggeri, A., Rocha, V., Masson, E., Labopin, M., Cunha, R., Absi, L., Boudifa, A., Coeffic, B., Devys, A., De Matteis, M., Dubois, V., Hanau, D., Hau, F., Jollet, I., Masson, D., Pedron, B., Perrier, P., Picard, C., Ramouneau-Pigot, A., Volt, F., Charron, D., Gluckman, E., Loiseau, P., 2013. Impact of donor-specific anti-HLA antibodies on graft failure and survival after reduced intensity contitioning-unrelated cord blood transplantation; a Eurocord, Société Francophone d'Histocompatibilité et d'Immunogénétique (SFHI) and Société Française de Greffe de Moelle et de Thérapie Cellulaire (SFGM-TC) analysis. Stem Cell Transpl. 98 (7), 1154–1160.

Schinstock, C.A., Mannon, R.B., Budde, K., Chong, A., Haas, M., Knechtle, S., Lefaucheur, C., Montgomery, R.A., Nickerson, P., Tullius, S.G., Ahn, C., Askar, M., Crespo, M., Chadban, S.J., Feng, S., Jordan, S.C., Man, K., Mengel, M., Morris, R.E., O'Doherty, I., Ozdemir, B.H., Seron, D., Tambur, A.R., Tanabe, K., Taupin, J.-L., O'Connell, P.J., 2020. Recommended treatment for antibody-mediated rejection after kidney transplantation: the 2019 expert consensus from the transplant society working group. Transplantation 104 (5), 911–922.

Selwood, N., Hedges, A., 1978. Transplantation Antigens – A Study in Serological Data Analysis. John Wiley & Sons, New York, pp. 22–43 (Chapter 3).

Sullivan, H.C., Gebel, H.M., Bray, R.A., 2017. Understanding solid-phase HLA antibody assays and the value of MFI. Hum. Immunol. 78, 471–480.

Sullivan, H.C., Dean, C.L., Liwski, R.S., Biswas, S., Goodman, A.L., Krummey, S., Gebel, H.M., Bray, R.A., 2018. (F)Utility of the physical crossmatch for living donor evaluations in the age of the virtual crossmatch. Hum. Immunol. 79 (10), 711–715.

Sunderland, C.A., Naiem, M., Mason, D.Y., Redman, C.W.G., Stirrat, G.M., 1981. The expression of major histocompatibility antigens by human chorionic villi. J. Reprod. Immunol. 3, 323–331.

Tait, B.D., 2016. Detection of HLA antibodies in organ transplant recipients – triumphs and challenges of the solid phase bead assay. Front. Immunol. 7, 570.

Tait, B.D., Süsal, C., Gebel, H.M., Nickerson, P.W., Zachary, A.S., Claas, F.H.J., Reed, E.F., Bray, R.A., Campbell, P., Chapman, J.R., Coates, P.T., Colvin, R.B., Cozzi, E., Doxiadis, I.I.N., Fuggle, S.V., Gill, J., Glotz, D., Lachmann, N., Mohanakumar, T., Suciu-Foca, N., Sumitran-Holgersson, S., Tanabe, K., Taylor, C.J., Tyan, D.B., Webster, A., Zeevi, A., Opelz, G., 2013. Consensus guidelines on the testing and clinical management issues associated with HLA and non-HLA antibodies in transplantation. Transplantation 95 (1), 19–47.

Tambur, A.R., Herrera, N.D., Haarberg, K.M.K., Cusick, M.F., Gordon, R.A., Leventhal, J.R., Friedewald, J.J., Glotz, D., 2015. Assessing antibody strength: comparison of MFI, C1q, and titer information. Am. J. Transplant. 15 (9), 2421–2430.

Voller, A., Bidwell, D.E., 1976. Enzyme-immunoassays for antibodies in measles, cytomegalovirus infections and after rubella vaccination. Br. J. Exp. Pathol. 57, 243–247.

Weissenbacher, A., Hautz, T., Zelger, B., Zelger, B.G., Mayr, V., Bradnacher, G., Pratschke, J., Schneeberger, S., 2014. Antibody-mediated rejection in hand transplantation. Transpl. Int. 27, e13–e17.

Williams, D., Rao, M., Carey, P., 2012. Impact of a transfusion-related acute lung injury reduction strategy on apheresis platelet collections. J. Clin. Apher. 27, 205–211.
Wozniak, L.J., Hickey, M.J., Venick, R.S., Vargas, J.H., Farmer, D.G., Busuttil, R.W., McDiarmid, S.V., Reed, E.F., 2015. Donor-specific HLA antibodies are associated with late allograft dysfunction after pediatric liver transplantation. Transplantation 99 (7), 1416–1422.
Zar, J.H., 1974. Biostatistical Analysis. Prentice-Hall, Englewood Cliffs, NJ, pp. 22–26.

CHAPTER 6

HLA typing by serologic methods

Contents

Bullet points:

1. Antibody-based methods detect human leukocyte antigens (HLAs) expressed on the target cell surface.
2. Antibody-based methods detect epitopes on the HLAs which may identify individual HLAs or be shared across multiple antigens, antigen groups, and HLA loci.
3. Antibody-based methods have been replaced by molecular methods for typing.

Human leukocyte antigen (HLA) typing (assignment of HLA-A, -B, -C, -DR, -DQ, and -DP) by serological methods has been replaced by molecular methods in clinical use. Serological HLA typing will be discussed for its historical importance and to put the current and evolving use of molecular methods into context. There is still significant value in determination of the HLAs expressed on the cell surface, particularly, because it is the antigens that are recognized by antibody directed against HLA. As will be discussed, the limitations of serological typing and the robust consistency of the DNA-based molecular methods have driven clinical HLA typing away from serological methods to molecular methods.

Serological HLA typing relies on antibody binding to the expressed HLAs on the surface of target cells, usually lymphocytes, as described in Chapters 4 and 5. Serologic typing can be described as a panel of crossmatches between sera with defined antibody reactivities against the unknown antigen profile of the target cells. The microlymphocytoxicity method became the standard method for serological HLA typing as it had for crossmatching and antibody detection and identification. By my count, there are 139 recognized HLA specificities today compared with the continually increasing thousands of alleles recognized (as previously discussed). In 1968, when only two HLAs were formally recognized, typing

HLA from Benchtop to Bedside
ISBN 978-0-12-823976-6
https://doi.org/10.1016/B978-0-12-823976-6.00006-8

was much less complicated than in 1991 when the last WHO antigen designations were assigned (Bodmer et al., 1992).

Prior to the availability of commercial HLA typing trays, laboratories obtained typing sera from colleagues, the National Institutes of Health, and by screening volunteers and blood donors for HLA antibodies. Many HLA labs used staff members and hospital staff members as "walking donors" for lymphocytes used for screening sera. Even after commercial typing trays became widely available in the 1980s (Dr. Paul Terasaki's company, One Lambda, was incorporated in 1984), most large HLA laboratories continued to use a mixture of "local" typing trays, exchanged and locally identified sera, and purchased typing sera, and prefilled typing trays. One of the laboratories I directed screened the plasma from multiparous blood donors for reagent-quality HLA antibodies in a partnership with the American Red Cross Blood Services (see reference for similar program in Canada (Mervart and Moore, 1983)).

A significant advantage of the commercial typing trays was access to sera with specificities that were unusual in the local population. Likewise, a disadvantage of the commercial typing trays and reagents was that specificities found in people of African, Native American, and Pacific Rim populations were underrepresented until special trays were created to cover antigens underrepresented in predominantly Caucasian (European decent) populations.

The most significant advantage of serological typing is that it detects the expressed HLA on the surface of the target lymphocytes (likewise, a null mutation will result in a nonexpressed antigen). The disadvantages are numerous:

1. Lymphocytes are surrogates for the organs transplanted: the HLA expression of the kidney, for example, is not typed in preparation for transplant.
2. Shared epitopes between HLAs from a locus and between antigens from different loci result in crossreactive antigen patterns and increased likelihood of mistakes in assignments, particularly for complex families of antigens such as the HLA-B15 group.
3. HLA Class II serological typing requires use of B lymphocytes which are in lower numbers in peripheral blood and more difficult to maintain viable than T lymphocytes.
4. Serological typing requires separation of T and B lymphocytes prior to typing to identify reactivity to Class I and Class II antigens. Some

fluorescent antibody methods were introduced to improve Class II typing but required fluorescence microscopy.
5. Human sera with antibody to HLAs are most commonly polyspecific, and both antibody specificity and concentration of antibody are likely to change over time in a donor.
6. Epitopes recognized by monoclonal antibodies may result in reaction patterns that differ from the polyclonal antibody response seen in human sera. These epitope reactivities are similar to the "splits" of HLAs that were recognized by polyclonal typing reagents.
7. Screening donors is laborious, and volunteer donors may choose to not return. An excellent monospecific reagent may be "lost."

The introduction of molecular typing methods improved the reliability and consistency of HLA typing both within and between laboratories as demonstrated in proficiency testing results, particularly in comparison to antibody-based typing methods (Duquesnoy and Marrari, 1994). The proficiency testing improvements continued as did the deceased donor typing discrepancy rates between laboratories reported by the United Network for Organ Sharing (UNOS) (Colombe and Lou, 1997). The College of American Pathologists discontinued proficiency testing for serological HLA typing (HLAS survey) in 2017 due to lack of participants and the replacement of antibody-based methods by molecular typing. Serologic typing is rarely used except to clarify unusual HLA types or to assist in the identification of novel null or low expression alleles (Mark Fung, MD, personal communication).

References

Bodmer, J.G., Marsh, S.G.E., Albert, E.D., Bodmer, W.F., Dupont, B., Erlich, H.A., Mach, B., Mayr, W.R., Parham, P., Sasazuki, T., Schreuder, G.M.Th., Strominger, J.L., Svejgaard, A., Terasaki, P.I., 1992. Nomenclature for factors of the HLA system, 1991. Tissue Antigens 39, 161–173.

Bozón, M.V., Delgado, J.C., Selvakumar, A., Clavijo, O.P., Salazar, M., Ohashi, M., Alosco, S.M., Russell, J., Yu, N., Dupont, B., Yunis, E.J., 1997. Error rate for HLA-B antigen assignment by serology: implications for proficiency testing and utilization of DNA-based typing methods. Tissue Antigens 50 (4), 387–394.

Bozón, M.V., Delgado, J.C., Turbay, D., Salazar, M., Granja, C.B., Alosco, S.M., Dupont, B., Yunis, E.J., 1996. Comparison of HLA-A antigen typing by serology with two polymerase chain reaction based DNA typing methods: implications for proficiency testing. Tissue Antigens 47, 512–518.

Colombe, B.W., Lou, C.D., 1997. An analysis of discrepancies in HLA phenotypes of cadaver donors in the united network for organ sharing registry. Transplantation 64 (12), 1786–1794.

Duquesnoy, R.J., Marrari, M., 1994. Progress report on the ASHI/CAP proficiency survey program in histocompatibility testing II. HLA-DR,DQ serologic typing, antibody identification, and B-cell crossmatching. Hum. Immunol. 39, 96–105.
Mervart, H., Moore, B.P., 1983. HLA typing in Canada: production and free distribution of tissue-typing trays. Can. Med. Assoc. J. 129 (7), 689–690.
Mytilineos, J., Scherer, S., Hansen, B., Gaweco, A., Opelz, G., 1990. RFLP-DR beta and serological HLA-DR typing of 200 kidney recipients and 1000 controls. Transplant. Proc. 22 (4), 1911–1912.
Opelz, G., Mytilineos, J., Scherer, S., Dunckley, H., Trejaut, J., Chapman, J., Middleton, D., Savage, D., Fischer, G.F., Bignon, J.D., Bensa, J.C., Albert, E., Noreen, H., 1991. Survival of DNA HLA-DR typed and matched cadaver kidney transplants. Lancet 338 (8765), 461–463.

CHAPTER 7

HLA typing by molecular methods

Contents

Bullet points:

1. Molecular methods detect the nucleotide sequences of the human leukocyte antigen (HLA) genes.
2. Molecular methods must discriminate between HLA loci, including noncoding loci of the major histocompatibility complex (MHC).
3. Polymorphisms in HLA alleles include nucleotide substitutions that result in amino acid substitutions, null alleles, expression changes, and nucleotide changes that do not change the amino acid sequence ("silent alleles").
4. Current clinical molecular typing methods utilize the amplification power of the polymerase chain reaction.
5. Sequencing has become the gold standard against which clinical typing is validated and is essential for high resolution HLA typing.
6. Sequence-based typing of HLA alleles is primarily directed at the hypervariable regions of the HLA gene loci which correspond to exons 2 and 3 for HLA Class I and exon 2 for HLA Class II genes.
7. Typing ambiguities may be resolved by targeted sequencing of polymorphism in exons beyond the peptide-binding regions.

HLA from Benchtop to Bedside
ISBN 978-0-12-823976-6
https://doi.org/10.1016/B978-0-12-823976-6.00007-X

8. Massively parallel sequencing (Chapter 8), also known as next generation sequencing (NGS), is revolutionizing DNA sequencing and minimizes many of the limitations of Sanger sequencing.

The standard of care for transplantation and the state of the art for human leukocyte antigen (HLA) typing is DNA sequence—based molecular typing. The most commonly used methods in the United States (in 2018) are PCR-SSP (polymerase chain reaction-sequence specific amplification), bead-based reverse sequence specific oligonucleotide binding, microcapillary Sanger sequencing, and the various forms of "next generation sequencing" (NGS; massively parallel nucleotide sequencing). Real-time DNA PCR methods (RT-PCR or qPCR) are being successfully used in laboratories needing rapid molecular typing giving low resolution (antigen level) DNA typing.

Molecular methods for clinical HLA typing were adopted because the technology became competitive with existing serologic methods after the introduction of the PCR for DNA amplification (see below) and, significantly, because of the serious technical issues in typing HLA Class II by serological methods. Challenges and limitations for serological HLA Class II typing included complicated isolation procedures for B lymphocytes, extensive crossreactivity between Class II antigens and typing failures (Bugawan et al., 1988). In contrast to lymphocytotoxicity typing, there is no need for viable cells with molecular methods. Nuclear DNA can be isolated from peripheral blood, lymph nodes, frozen cells, cheek swabs, and processed surgical biopsies.

Molecular HLA typing "before PCR"

Sequencing

Although scientists studying HLA were "early adopters" of molecular methods for HLA typing, molecular sequencing methods predate the methods adopted by HLA typing laboratories, and DNA sequencing methods (and the sequences obtained) made the subsequently developed typing methods possible. Two successful DNA sequencing methods were published in 1977. The first is presented for historical completeness (Maxam and Gilbert, 1977). The second method (Sanger et al., 1977) is still the most widely used sequencing method, only now being challenged by massively parallel sequencing, also called NGS, which will be discussed in Chapter 8.

Although the method has been described as the "first-generation" sequencing method in a review by Heather and Chain (2016), the method of Maxam and Gilbert was eclipsed by the methods that were developed by Sanger et al. (see following discussion). The Maxam and Gilbert method is a chemical cleavage method based on susceptibility of specific bases to chemical degradation. A multistep procedure, chemical treatments are selected to degrade 5′ and 3′ ^{32}P labeled DNA fragments based on the presence of a purine or pyrimidine. Dimethyl sulfate is used to methylate and destabilize the purines which are then further fragmented in alkali at 90°C. Resolution of guanine and adenine is achieved by a separate treatment in which treatment with dilute acid preferentially releases adenines. Hydrazine is used to cleave thymine and cytosine. The product is cleaved with piperidine. Cytosine is differentiated from thymine by suppression of the reaction of hydrazine and thymine by 2 M NaCl in a separate reaction. The reaction products of the four treatments are separated in parallel polyacrylamide gel electrophoresis and radiographs produced from which the sequences can be read. The Maxam and Gilbert sequencing method is robust, and the sequences are readily resolved but limited to relatively short DNA sequences (~100 bases). The method was very successful for sequencing relatively short DNA sequences from bacteriophages and short sequences from clonal libraries.

The highly successful Sanger sequencing method (Sanger et al., 1977) revolutionized molecular biology and made molecular genetics a reality. The method is elegant in its simplicity; isolated fragments of DNA are "copied" by a DNA polymerase in the presence of dideoxynucleotides (ddNTPs) or arabinosyl (ara) nucleotides. (Eventually, only ddATP, ddTTP, ddCTP, and ddGTP were utilized). The ddNTPs are incorporated into the growing sequence, but no further nucleotides can be added because the ddNTPs lack the 3′-OH group necessary for chain extension. Each chain-terminating nucleotide was run in parallel, and the radiolabeled products separated by electrophoresis in acrylamide gels. The four replication products are "read" from the smallest (fastest moving) product which corresponds to the earliest termination; each termination indicates which ddNTP was next incorporated. The original procedure used ^{32}P labeled DNA and autoradiographs of the electrophoretic gels and allowed for sequencing of isolated DNA up to 300 base pairs (bp) in length, although lengths shorter than 200 bp were more reliable. The use of radionucleotides in Sanger sequencing was made obsolete with the introduction of fluorescently labeled ddNTPs (Prober et al., 1987).

Early Sanger sequencing methods utilized cloning of fragmented DNA into plasmids for isolation and expansion prior to sequencing. Advances in Sanger sequencing have included PCR amplification of target DNA for sequencing, capillary electrophoresis on reusable columns and simultaneous detection of the incorporated fluorescent ddNTPs as the sequencing products pass the fluorescence detectors. Single reaction stream detection reduced the difficulty in arranging the incorporation sequences of the nucleotides (a significant problem when each nucleotide was evaluated in a separate electrophoretic column, even when run in parallel "channels" in a slab gel). Modern sequencing software systems facilitate analysis of the electrophoretic histograms (Fig. 24). Capillary systems with simultaneous detection of the four labeled nucleotides are standard in modern automated Sanger sequencers.

Sequence-based typing methods for HLA alleles are directed at the most informative regions of the alleles, primarily exons 2 and 3 for HLA Class I and exon 2 of HLA Class II alleles. The most polymorphic regions were recognized as being near the N-terminus of the proteins as early as 1983 (Long et al., 1983). As more alleles were identified by sequence-based typing methods and continue to be identified, nucleotide polymorphisms in other exons may be used to distinguish the alleles. Some clinical HLA typing strategies may employ more than one method to resolve HLA alleles (e.g., Sanger sequencing plus PCR-SSP targeting one or more single nucleotide polymorphisms in other exons) (Delfino et al., 1998).

Historical note: It should be noted that the revolutionary Sanger sequencing method was derived from another "primed" sequencing method called "plus minus sequencing" reported by Sanger and Coulson in 1975. For the perspective of 4 decades later, the method seems incredibly laborious but it allowed sequences to be obtained from native DNA. The priming sequences were either synthetic oligonucleotides or fragments from the target DNA. For historical purposes, the "minus" system synthesized DNA from the template in the absence of one of the four nucleotides (one of the three is labeled with ^{32}P). The polymerase would add nucleotides until reaching the corresponding base of the missing nucleotide and stop. The amplified oligonucleotides were then isolated by gel electrophoresis, and the sizes of the products indicate when the synthesis stopped. The process was repeated with each of the four nucleotides. The "plus" system used the exonuclease activity of a bacterial DNA polymerase in the presence of only one of the four nucleotides. The exonuclease would proceed until reaching the corresponding nucleotide in the target oligonucleotide and stop. Again, there would be separation by gel electrophoresis and repetition with addition of each of the other four nucleotides. The autoradiographs of the electrophoretic products were analyzed for the product sizes which correspond to the locations of each of the nucleotides in the sequence. Product size corresponding to incorporation of individual nucleotides became the basis for Sanger sequencing until individual nucleotide incorporation could be identified by fluorescent labeling.

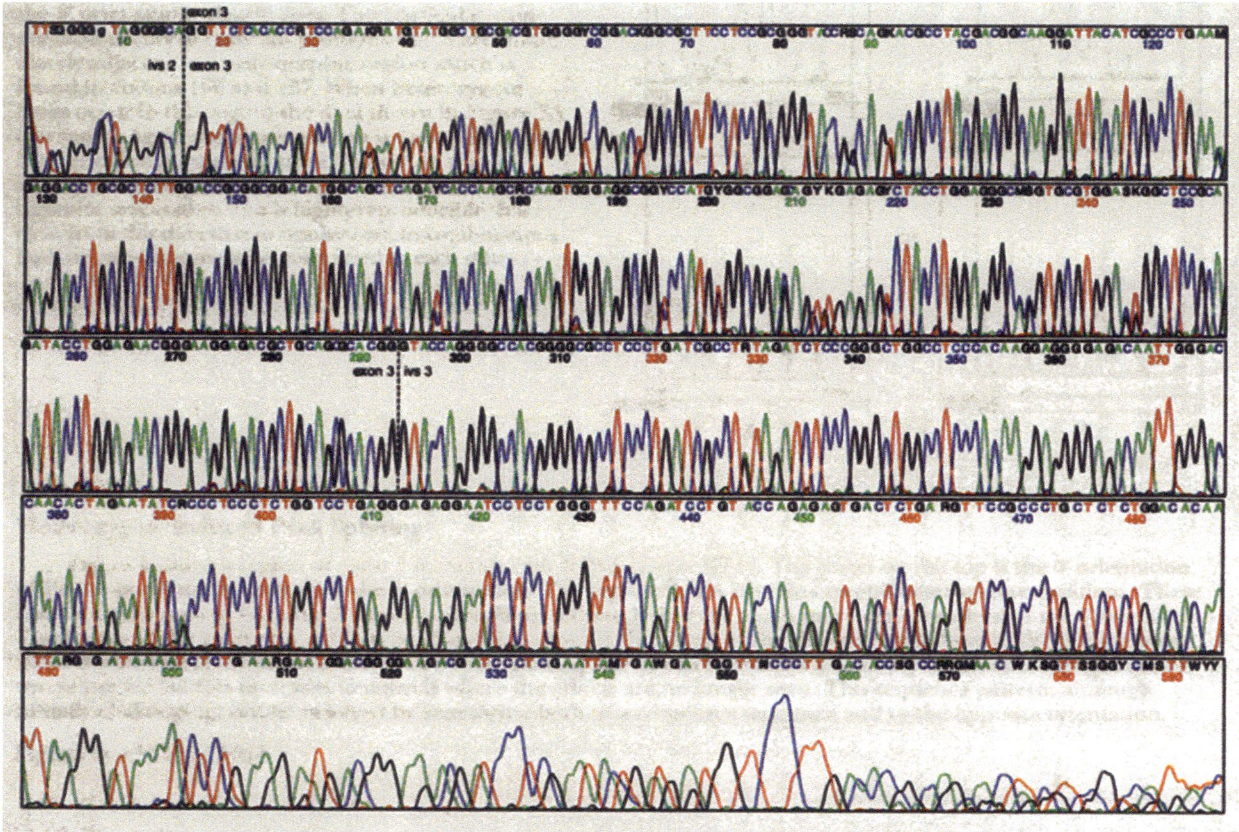

Figure 24 Example of a Sanger sequencing histogram. The smallest sequencing products (earliest incorporation of ddNTPs terminating the synthesis) are in the upper left corner. Sanger sequencing is of lowest quality in the initial base reads and with the longest reads (generally > 700 bases). Note the quality and sharpness of the peaks from about position 15 through 480 in this sequencing histogram.

Oligonucleotide probing and endonuclease cutting

Two related methods of HLA typing using molecular methods were introduced to research and clinical laboratories in the late 1980s: sequence-specific oligonucleotide probes (SSOP) and restriction fragment length polymorphism (RFLP). Both methods relied on the foundational work of Edwin Southern and the DNA transfer ("blotting") technique which is commonly referred to as "Southern blotting" (Southern, 1975). The foundational Southern technique has been eclipsed in most research and clinical use by methods using PCR (Saiki et al., 1985).

SSO and SSOP: Sequence-specific oligonucleotide probes (also known as ASO: allele specific oligonucleotide) were introduced in 1985 with the availability of exon-specific probes for sequences of the HLA-DRB region (Bidwell et al., 1986). As first described, SSOP used restriction fragmented genomic DNA that was separated by electrophoresis, transferred to nylon membranes by capillary blotting, and probed with radionucleotide-labeled oligonucleotides (Southern, 1975). Although a powerful research tool, the assay in its early forms was of limited use clinically; the assay consumed significant amounts of DNA, required radiolabeling of the oligonucleotide probes (usually ^{32}P), and required multiple days of exposure of X-ray film to the probed membranes to obtain the results (or learn that the assay failed and you were back to square one). Probe binding patterns were compared to known typing for interpretation, and previously unknown types were often identified by new patterns of binding of the radioactive probes.

The principles of SSOP typing are based on the specificity of oligonucleotide binding to target nucleotide sequences. The "stringency" or specificity of the probe binding to the target sequence is determined by physical chemistry constraints: complimentarity of the sequences, length of the probe, binding energy between nucleotides, "nearest neighbor" effect, temperature and ionic conditions of the reaction, and wash fluids (Breslauer et al., 1986). Wash conditions for SSOP assays (primarily temperature and salt concentrations) are selected so that single nucleotide mismatches between the probe and target sequence will prevent binding before the detection step is initiated. More stringent conditions (usually in the wash steps) result in more specificity of the sequence detection. Although calculators based on the Breslauer et al. (1986) formulas are readily available (see for example, http://biotools.nubic.northwestern.edu/OligoCalc.html) (Kibbe, 2007), optimization of conditions requires some experimentation such as adjusting magnesium or manganese ion concentrations.

Advances in SSO typing came quickly with expanded sequence libraries for the known HLAs allowing for the synthesis of new oligonucleotide probes, introduction of instruments to perform standardized, semi-automated drop blots of DNA onto nylon membranes, and the introduction of colorimetric and chemiluminescent detection methods which eliminated the use of radionucleotides and dramatically shortened assay times (Pollard-Knight et al., 1990).

Restriction fragment length polymorphism

Like SSO, one of the first molecular methods for HLA typing adopted in many clinical laboratories was RFLP applied to HLA-DR typing (Carlsson et al., 1987; Mytilineos et al., 1990). The complicated method involved isolation of genomic DNA from 10 to 15 mL of EDTA anticoagulated peripheral blood, splenic tissue, or frozen lymphocytes. The isolated DNA was digested with restriction endonucleases of known specificity and separated by agarose gel electrophoresis. After electrophoresis, the separated DNA fragments were transferred to nitrocellulose membranes for hybridization ("Southern blotting" see above) and hybridized with ^{32}P labeled cDNA probes prepared from clonal libraries derived from coding sequences from HLA-DRB and -DQB (in research citations HLA-DQA and -DPB were frequently included). As with SSO, hybridization patterns were compared to patterns identified with well-characterized serologically defined HLA types.

Polymerase chain reaction and the molecular revolution in clinical medicine

The principles of PCR are deceptively simple and resulted in the 1993 Nobel Prize in Chemistry for Kary B Mullis, Ph.D., and fundamental changes and expansion of the use of DNA sequence analysis in research and clinical laboratory science. The PCR uses oligonucleotide primer sequences to initiate nucleotide replication in the presence of excess nucleotides (deoxynucleoside triphosphates), a DNA polymerase, and divalent cations (typically Mg^{2+}) in a buffer solution. An early and significant advance in PCR technology was the use of thermostable DNA polymerase to eliminate the need for adding new polymerase prior to each amplification sequence (Saiki et al., 1988). The primer sequences are selected to "bracket" the sequence of nucleotides to be amplified (either from genomic DNA or cloned sequences from isolated DNA). The native, double-stranded DNA is heated to unwind and separate the complementary strands, then the temperature is lowered to allow the 5′ and 3′ primer oligonucleotides to bind. The binding or annealing temperature is critical because it contributes to the specificity of the primer binding by reducing mismatching between the primer sequences and the target DNA sequence. The temperature must be optimized for each primer pair, commonly 3–5°C below the melting temperature (Tm) of the primer oligonucleotide. After the brief annealing time, the temperature is raised to the optimal temperature of the DNA polymerase, and the free bases are added to the single strands (elongation phase), doubling the number of DNA strands containing the sequences found between the primer sequences. The process is repeated and the original DNA **and** the copies are duplicated in an exponential increase in the target sequence (Fig. 25). The introduction of programmable thermocyclers allowed for automation of the temperature cycling and greater consistency of the reaction once parameters are set.

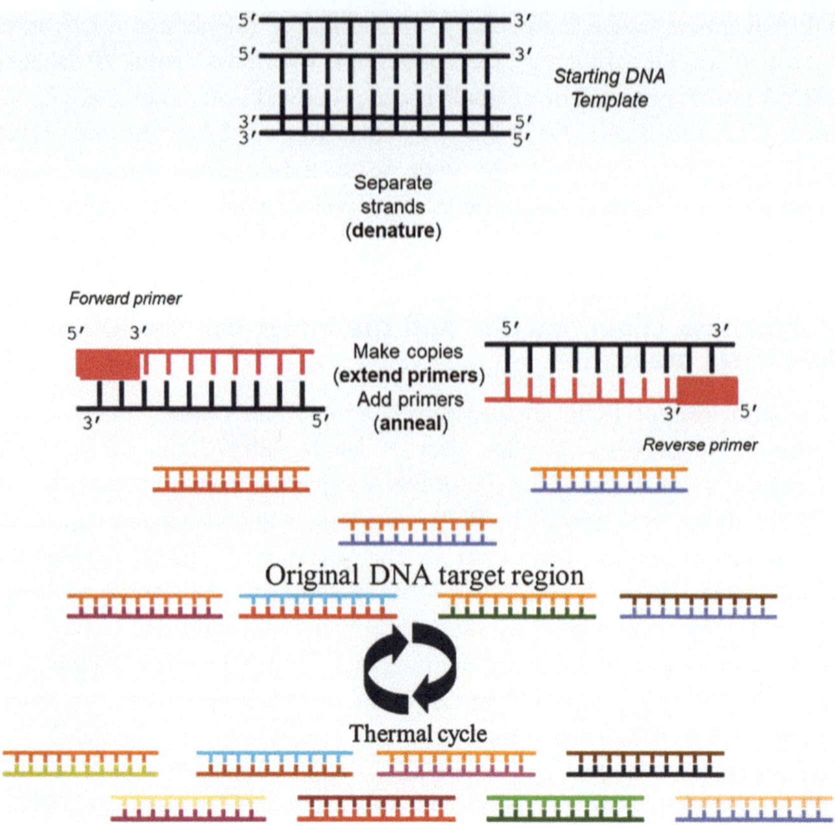

Figure 25 Cartoon of polymerase chain reaction (PCR) amplification of target DNA.

In theory, at 100% efficiency, 32 cycles of amplification result in 1.07 billion copies of the target sequence. Actual efficiencies depend on the accuracy of the DNA polymerase, quality of the DNA substrate, selection of optimal primer sequences, and optimization versus compromises in each part of the thermocycler program (Lorenz, 2012).

The introduction of the PCR technology to HLA typing was revolutionary (Erlich, 2012). PCR is critical to essentially all molecular HLA typing methods in clinical use today.

Polymerase chain reaction sequence specific primers

An early application of the PCR for HLA typing was the amplification of target sequences using combinations of 5′ and 3′ primers to amplify families of alleles and to discriminate between alleles by selecting primers specific for single nucleotide differences between alleles (Olerup and Zetterquist, 1991; Zetterquist and Olerup, 1992). Selecting primers is challenging for HLA typing because of the conservation of sequences between HLA alleles within an allelic family, sequence homology between expressed HLA loci, and, not surprisingly, between expressed loci and the pseudogenes in the major histocompatibility complex. Olerup and Zetterquist addressed the specificity issues by selecting 5′ primers that would be common to the second exon of allele groups within HLA-DRB1 and 3′ primers that would distinguish between alleles, either alone or in combination with the amplification pattern of other 3′ primers. The specificity of the amplification is determined by the amplification conditions and confirmed by the size of the amplification product detected in agarose gel electrophoresis (Fig. 26). Visualization of the amplification products is by an intercalating dye (e.g., ethidium bromide or cyanine dyes) examined with UV illumination. Estimation of the size of the product is by comparison with a "molecular weight ladder" composed of oligonucleotides of known size (number of base pairs) that are electrophoresed in lanes adjacent to the unknowns. Controls for amplification (such as beta-globulin genes) are included in the amplification mixture, and a DNA-free sample is included to detect contamination of reagents.

Commercial PCR-SSP kits have expanded on the original designs of Olerup and Zetterquist to cover the HLA Class I and II loci for solid organ transplant and supplemental typing for discriminating ambiguities that occur with most high resolution sequence-based HLA typing methods (see the section on Sanger sequencing for discussion of typing ambiguities).

Figure 26 Cartoon of PCR-SSP gel. Internal control bands are not shown for clarity.

PCR-SSP became widely adopted for clinical typing because it was competitive with serological methods both in cost and the assay time (suitable for solid organ donor typing, for example) and because the results were more accurate than serological typing, particularly for HLA Class II typing. The simplicity and rapidity of the PCR-SSP method has also been adapted for single or multiple locus typing for disease association testing such as HLA-B27.

The dark rectangles across the top of Fig. 26 represent sample wells for the electrophoresis gel. Column A is representative of a molecular weight "ladder" using oligonucleotides of known molecular weight for assessment that the PCR-SSP products are of the expected size (number of base pairs in length, smaller oligonucleotides move more rapidly toward the end of the gel). Columns B, C, and E represent the expected electrophoretic migration pattern for a positive amplification. The first "band" in columns B, D, D, and E would be an internal amplification control product such as beta-globulin. The lack of a second band in Column D indicates the absence of the allele detected by the primers. A positive control band (not shown in the cartoon) demonstrates that sample DNA was present, and the conditions were adequate for amplification.

PCR-RFLP

PCR-RFLP combines the discriminating power of the RFLP technique with the power of PCR amplification of specific target sequences within HLA loci. Previously identified nucleotide sequences, usually within the polymorphic peptide binding regions defined by the N-terminal regions distal to the cytoplasmic region of the HLA molecule, are searched for known restriction endonuclease cutting sites that can be used to identify allelic differences. Then, using PCR primers based on previously identified sequences in HLA loci, locus-specific amplifications are performed (see PCR-SSP, above). The amplified DNA products are then incubated with selected endonucleases, and the final products are separated by electrophoresis. The resulting patterns of bands (which can be compared to a molecular weight standard "ladder") are then compared to the expected pattern for known alleles. When a restriction endonuclease cutting site is present, the electrophoresis pattern will show lower weight (smaller) fragments of DNA instead of the larger, uncut, initial product (Fig. 27). When two alleles are amplified by the same primers but only one has the endonuclease cutting site, a pattern of the larger (uncut) and smaller (cut)

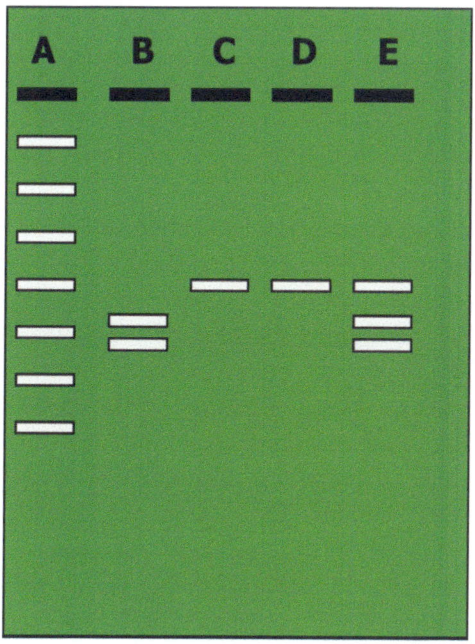

Figure 27 Cartoon of PCR-RFLP gel. Internal control bands are not shown for clarity.

bands is observed (Maeda et al., 1990). As identified by Mytilineos et al. (1990), RFLP significantly improved the reliability of HLA-DR typing when compared to serological methods. Although a very powerful technique, PCR-RFLP has been eclipsed by newer sequence-based molecular typing methods for clinical HLA typing.

The format of the PCR-RFLP gel in Fig. 27 is similar to that shown in Fig. 26 for PCR-SSP. The dark rectangles across the top of Fig. 27 represent sample wells for the electrophoresis gel. Column A is representative of a molecular weight "ladder" using oligonucleotides of known molecular weight for assessment of the PCR product size. Columns C and D show an expected pattern of PCR-SSP amplification products that lack the endonuclease cutting site which would identify the target allele. Column B shows a representative electrophoretic pattern for an allele that was amplified in the PCR-SSP reaction and has the restriction endonuclease cutting site; the allele in B is different than the allele(s) in columns C and D. Column E is an example of the electrophoretic pattern expected when the sample is from a person who has an allele lacking the cutting site (first band) and also has an allele with the target cutting site (second and third bands with electrophoretic pattern similar to column B).

PCR-rSSO

The most commonly used method for HLA typing in 2018 was microbead array assay in which oligonucleotide probes are cross-linked by a thymidine homopolymer tail to fluorescently labeled microbeads and patient DNA is amplified by locus-specific PCR with biotinylated primers. The amplified DNA is hybridized to the beads and washed under stringent conditions. Detection of binding of PCR-amplified patient DNA to the solid-phase bound oligonucleotide probes is by streptavidin-conjugated R-phycoerythrin (PE) binding to the biotinylated "tails" on the amplification primers (Fig. 28.). Each microbead is intrinsically labeled with a mixture of fluorescent dyes which allows the laser interrogation and identification of the bead and the specific oligonucleotide probe attached (Luminex®, Luminex Corporation, Austin, TX, USA).

The concept of "reverse SSO (sequence specific oligonucleotide probes)" originated as the "reverse dot blot" (Saiki et al., 1989). The rSSO concept was fully applied to HLA typing using the sequence-specific oligonucleotide probes (SSOP) crosslinked to nylon membranes to which were hybridized the locus-specific PCR products generated with

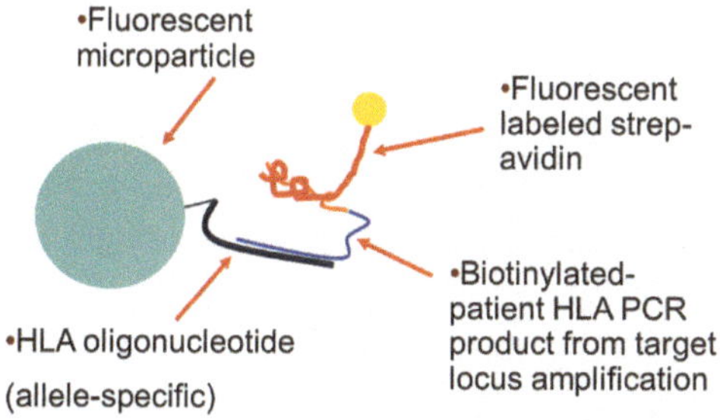

Figure 28 rSSO microbead assay for HLA typing.

biotinylated primers (Erlich et al., 1991). The method was later adapted to other solid supports such as plastic cards and, finally, to the fluorescently labeled microbeads now used in the commercial HLA rSSO typing kits for the Luminex® platform (Erlich, 2012).

A specific challenge of the rSSO technique is optimization of the solid-phase probes so that a single wash fluid can provide the appropriate and stringent conditions for all of the probe-product binding pairs. Some probes are lengthened, some are shortened, and the density of probe on the solid surface may be adjusted to "balance" the signal from the probe-product reaction pairs (Erlich et al., 1991).

Current (December 2018) rSSO HLA typing kits are the microbead LABType® SSO and SSO HD (Thermo Fisher Scientific, Inc, USA) and LIFECODES HLA Typing Kits—Rapid (Immucor, Inc, USA) for the Luminex® platform, and HISTO SPOT® SSO (BAG Health Care GmbH, Germany) which is a well-based microspot colorimetric assay using the MR.SPOT® Processor for detection and analysis (BAG Health Care GmbH, Germany). The HISTO SPOT® products are not available in the United States (December 2018). The LABType® rSSO method uses balanced amplifications and denaturization of the amplicons to produce single-strand DNA prior to hybridization. The LIFECODES HLA rSSO method uses an "asymmetric PCR" in which one primer is in excess resulting in the generation of some single-strand amplicons to bind to the oligonucleotide probes. Both methods perform the hybridization step at high temperature to ensure the specificity of the binding (Dunn, 2015).

In September 2018, there were over 19,600 HLA alleles (HLA-A, -B, -C, -DRB1, -DRB3, -DRB4, -DRB5, -DQB1, and -DPB1; http://hla.alleles.org/nomenclature/stats.html). As noted previously, most of the sequence variation in HLA alleles is in the first and second domains of the HLA Class I and second domain of Class II molecules. HLA typing strategies are directed to those distinguishing sequences. Significant variability exists in the distal sequences, and some common null and alternatively expressed alleles are identified by sequences within and beyond exon 4 in the alpha-3 domain. The number of oligonucleotide probes necessary to distinguish the common and well-documented (CWD) HLA alleles (http://igdawg.org/cwd.html) continues to increase creating interpretation challenges requiring sophisticated analytic programs which are, fortunately, provided by the HLA typing kit vendors. The CWD alleles are those types which HLA laboratories are expected (required by most accrediting agencies) to be able to distinguish, particularly in support of hematopoietic stem cell transplantation. Today, the technologist, supervisor, and laboratory director are primarily limited to determining mean fluorescence intensity (MFI) cutoffs and examining whether changing an individual bead or microdot from negative to positive (or reverse) changes a typing assignment by the computer program (Figs. 29 and 30, Luminex® interpretation histograms). Interpretation of SSO and rSSO patterns was simpler when there were fewer alleles identified and fewer primer combinations to interpret (Figs. 31 and 32).

High resolution typing ambiguities in molecular HLA typing

One of the most significant limitations of all of the molecular typing methods in clinical use, with the exception of single-strand sequencing methods such as massively parallel sequencing or next generation sequencing (NGS, to be discussed in Chapter 8), is the difficulty in resolving typing ambiguities caused by the simultaneous sequencing of both the maternally and paternally derived chromosomal DNA (e.g., heterozygous base pair combinations that are shared by two or more alleles). Ambiguities have three common causes: heterozygous nucleotide pairs that are obtained when there is more than one possible pair of alleles with the detected nucleotides; alleles that are defined by polymorphisms outside of the target region (e.g., exon 4); new alleles or incompletely sequenced alleles (Adams et al., 2004). The European Bioinformatics Group has a tool for evaluating ambiguous HLA allele combinations and a frequently updated listing of the ambiguities that arise from sequence-based typing using exons 2 and 3 for HLA Class I typing and exon 2 for HLA Class II typing (https://www.ebi.ac.uk/ipd/imgt/hla/ambig.html). As noted in the nomenclature section (Chapter 3), HLA alleles that share identical amino

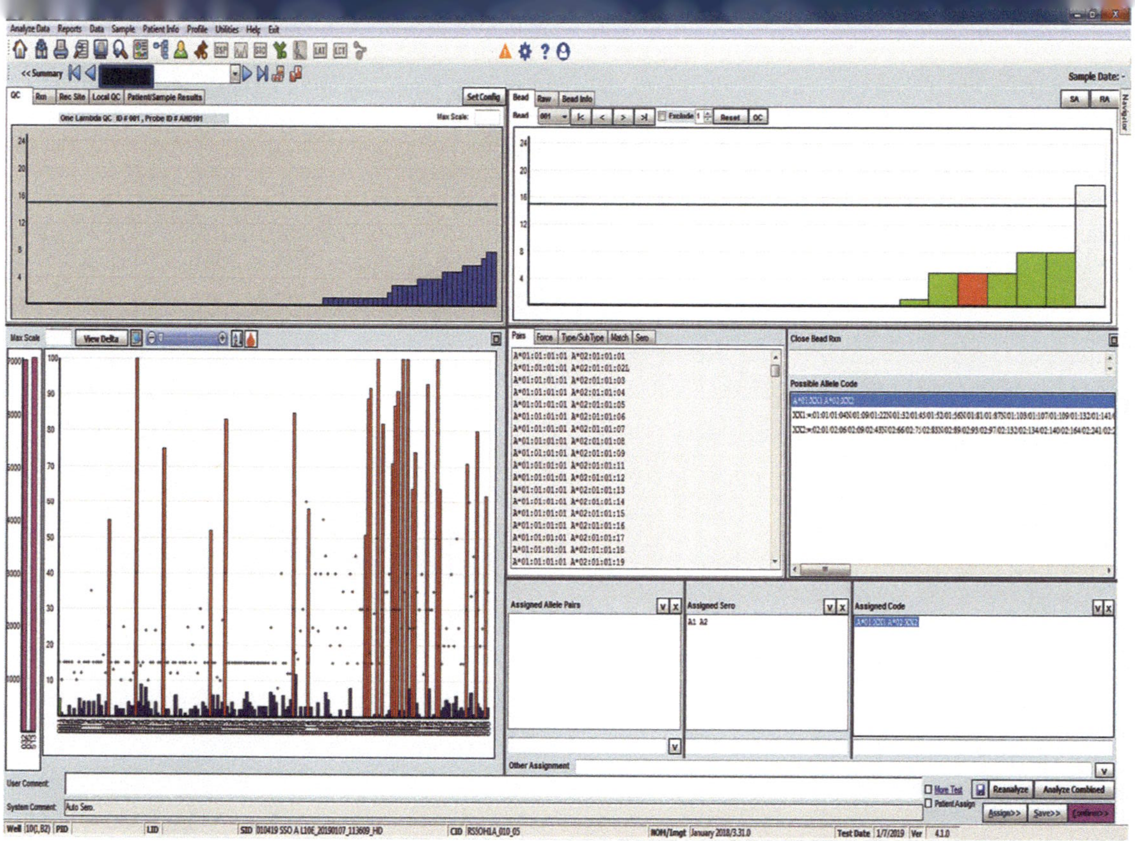

Figure 29 HLA-A*01:01:01G,*02:01:01G bead histogram from a Luminex®-based microparticle rSSO HLA typing assay analysis with proprietary Fusion™ software (LABType® SSO HD kit and Fusion™ software, Thermo Fisher Scientific, Waltham, MA). The center and center-right boxes list the possible alleles (ambiguous call assignment) that is expected with the common HLA-A*01:01,*02:01 typing. The ambiguities may be resolved and common null alleles excluded with supplemental testing such as PCR-SSP or Sanger sequencing if clinically indicated. *(Used with permission. Screenshot kindly provided by Sharon Skorupski, CHS, Henry Ford Hospital, Detroit.)*

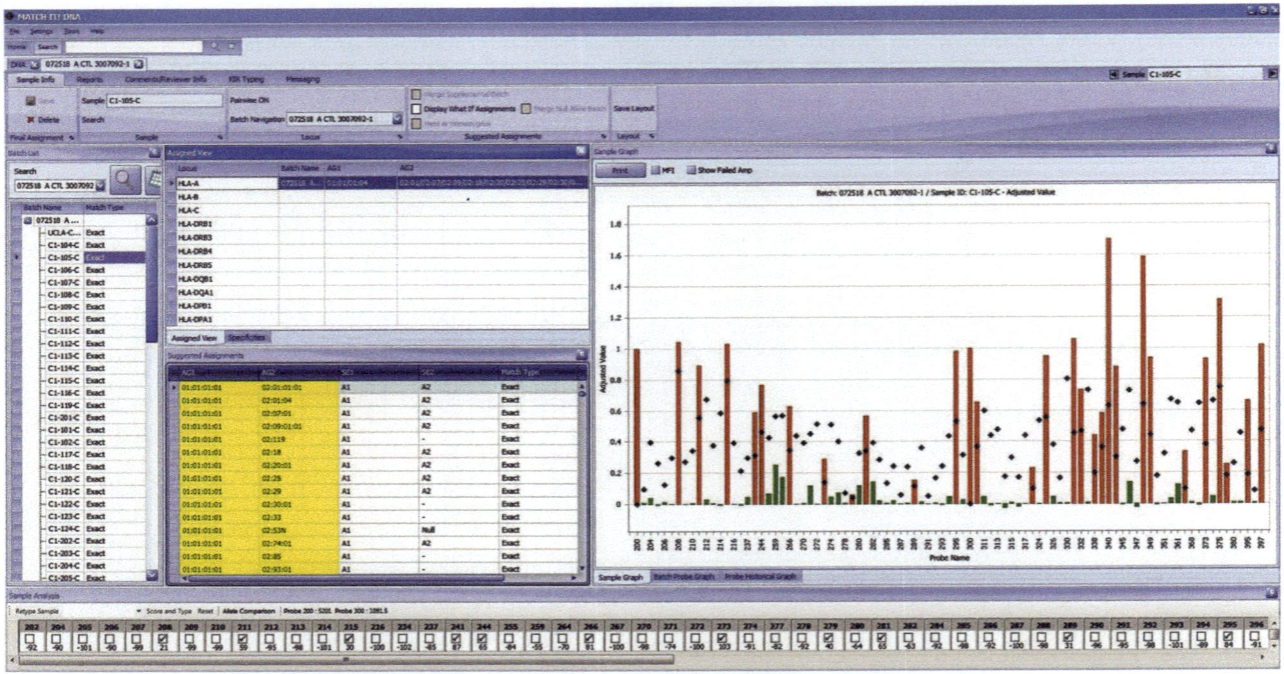

Figure 30 HLA-A*01:01:01G,*02:01:01G bead histogram from a Luminex®-based microparticle rSSO HLA typing assay analysis with proprietary Immucor Lifecodes® reagents and MATCH IT! DNA™ software (Immucor, Norcross, Georgia). The right histogram shows the amplification oligonucleotide product fluorescence intensity on each bead. The left center lower box lists the possible alleles (ambiguous call assignment) that is expected with the common HLA-A*01:01,*02:01 typing. The ambiguities may be resolved and common null alleles excluded with supplemental testing such as PCR-SSP or Sanger sequencing if clinically indicated. *(Used with permission. Screenshot kindly provided by Bryan Ray, Ph.D., Sr. Director R&D, Immucor, Waukesha.)*

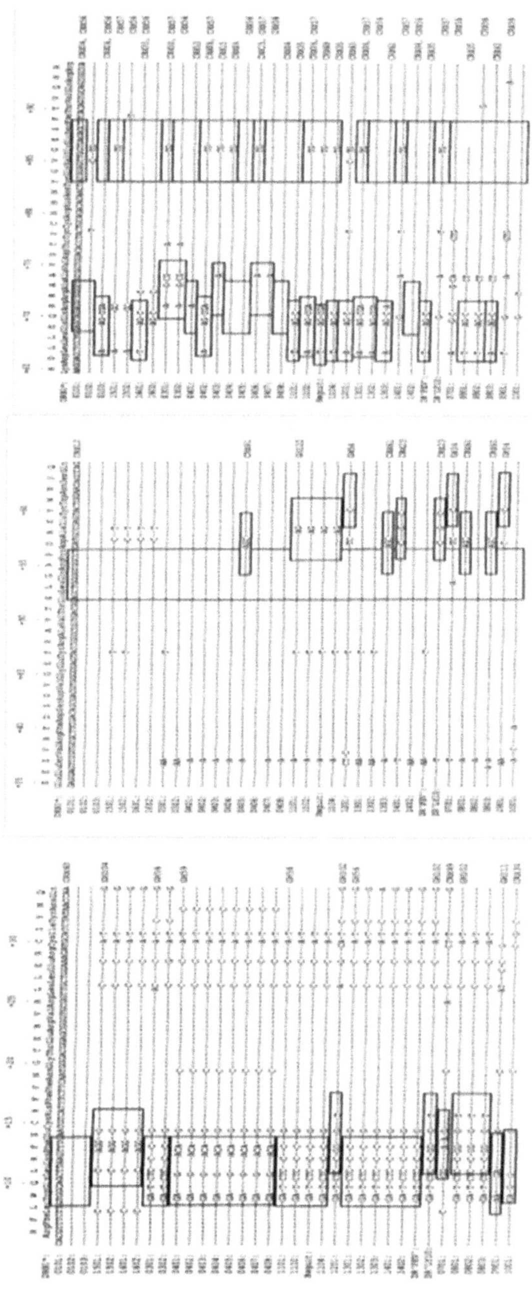

Figure 31 DRB1 locus DNA sequences and probe alignments from Erlich et al. (1991), using 26 probes and seven primers. *(Used with permission.)*

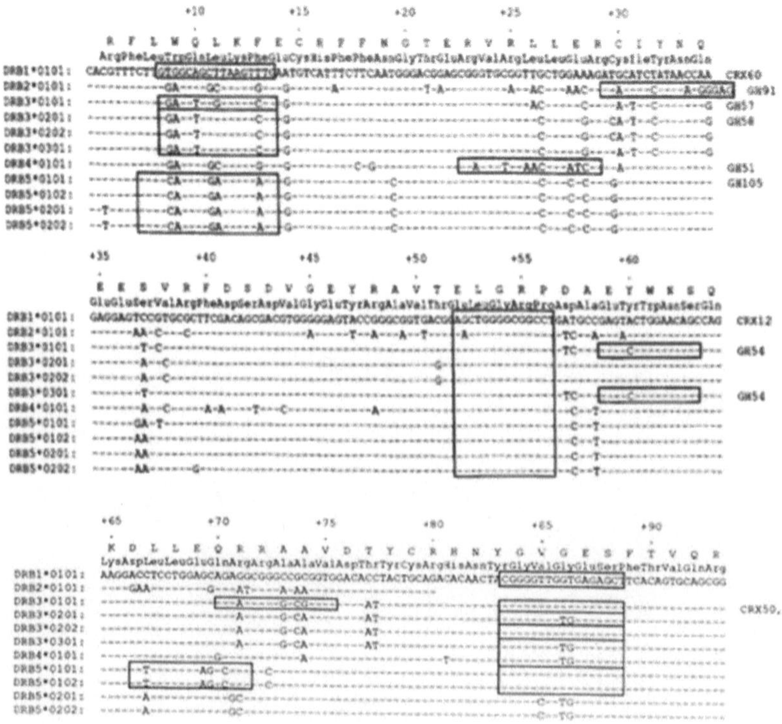

Figure 32 DRB1,2,3,4,5 locus DNA sequences and probe alignments from Erlich et al. (1991), using 26 probes and seven primers. *(Used with permission.)*

acid sequences through the peptide binding region may be designated by the letter "G" for the purpose of reporting (for example, A*02:01:01G). Although this simplifies reporting ambiguous allele combinations, G groups may include known null alleles that must be identified.

An excellent pictorial representation of a classic molecular typing phase ambiguity is presented by Madden and Chabot-Richards (2019), in a figure from a Sanger sequencing of DQB1*03 alleles. They note that the ambiguity can be resolved with allele-specific primers or NGS (to be discussed in the next chapter).

Expanded high resolution typing kits for rSSO include primers and probes to cover additional exons that may reduce ambiguities. Expanding the number of beads in the multiplex rSSO assays for the Luminex® platforms allows for more primer and oligonucleotide probe combinations as is seen with the new 500 bead scanners and assay systems designed to use the capability (Dunn, 2015).

The challenge of null alleles

Detection of nucleotide sequences associated with known HLA alleles does not mean that an HLA protein is expressed on the surface of that individual's nucleated cells. As noted in the earlier section on HLA nomenclature, abnormal expression of HLA alleles include null (N), low expression (L), secreted but not expressed on the cell surface (S), and questionable expression (Q; based on other alleles with similar nucleotide sequence changes). The cytoplasmic (C) and aberrant (A) classes of abnormal expression have not yet been assigned. The null and abnormal alleles are significant because expression of the HLA proteins is necessary for all of the known functions of this protein family (Elsner and Blasczyk, 2004). Low resolution HLA typing (defined as giving an HLA type roughly equivalent to a serologic or antigen type) and most methods of intermediate (allele group) and high resolution typing (individual alleles and excluding common null alleles) are usually obtained by targeting exon sequences in the oligopeptide-binding region of the HLA molecule (alpha-1 and alpha-2 domains of the Class I loci and alpha-2 and beta-2 domains of the Class II loci) (Nunes et al., 2011). PCR-based, oligonucleotide-binding based and Sanger sequencing methods can easily miss the more than 700 known null and abnormal alleles of the HLA Class I and Class II loci (http://hla.alleles.org/alleles/nulls.html). Clinically, this shortcoming is reduced (not eliminated) by the inclusion of accrediting organization standards requiring the typing for common and well-defined null HLA alleles (e.g., European Federation for Immunogenetics (EFI) Standards For Histocompatibility and Immunogenetics Testing, Version 7.0. F1.1.4.1.2; https://www.efi-web.org/fileadmin/user_upload/Website_documenten/EFI_Committees/Standards_Committee/2017-10-31_Standards_version_7.pdf).

Examination of the listing of null and abnormally expressed HLA alleles on the hla.alleles.orgalleles website finds that the most common cause is a point mutation that results in a premature stop codon. Other causes are deletions or insertions that result in a reading frameshift and a resultant premature stop codon, sequence changes that result in incorrect splicing of the exons, single nucleotide changes that result in structural abnormalities affecting disulphide bridges, beta-2-microglobulin binding or the insertion of the HLA protein into the cell membrane, or translation of introns that result in truncated proteins. Tools and downloadable data sets on the http://igdawg.org/cwd.html website provide a consensus listing of the CWD HLA alleles, null, alleles and allele groups that accredited HLA laboratories are expected to be able to identify.

Clinically, it is critical that common known null alleles are excluded when assigning HLA types, particularly in transplantation (this is not just a "regulatory or accreditation" issue). Abnormal or absence of expression will affect the immune recognition between the recipient and the donor cells or tissues. A simple example is a donor with a null allele at the HLA-DQB1 locus that would result in donor cells recognizing the "matched" recipient as a mismatch for one HLA-DQB1 allele and antigen. Similarly, if a solid organ recipient had an undetected null allele at the HLA-B locus, the patient would be likely to produce a de novo HLA antibody against the expected (but absent) HLA. Ironically, this would not have been an issue when typing was performed with serologic methods and viable lymphocytes.

rtPCR (aka qPCR, real-time PCR, and quantitative PCR)

Purified double-stranded DNA "melts," that is, separates base pair by base pair, at a temperature determined by a combination of physical effects from the guanine-cytosine (GC) content, the "nearest neighbor" effect on stability of the duplex chain, length, and ionic conditions of the solution, as discussed in the section on PCR-rSSO typing. The Tm of purified double-stranded DNA can be determined from the 50% change point in an absorbance curve at the 260 nm wavelength. The melt curve and Tm can also be determined by the release of a fluorescent dye that binds to double-stranded DNA (example: SYBR® Green, Bi-Rad Laboratories, Inc, Hercules, CA). The Tms of DNA amplified with specific primers can be used to detect sequence differences between products from allelic DNAs by comparing the detected Tm to a library of the Tms of known alleles amplified by the primer pair (Dunn, 2015). Alternative methods for rtPCR in HLA typing have been described including the use of a combination of group-specific primers and FRET-labeled informative probes (FRET: Fluorescence Resonance Energy Transfer) to identify alleles (Faner et al., 2006) or a combination of group-specific primers and TaqMan™ (Thermo Fisher Scientific, Waltham, MA) chemistries with allele-specific probes that fluoresce during the amplification allowing direct quantitation of product (Gersuk and Nepom, 2006). The probe methods are most advantageous for targeted HLA typing for specific allele groups (e.g., HLA-B27) or disease-risk or pharmacogenetic risks of specific alleles (e.g., HLA-B*57:01).

References

Adams, S.D., Barracchini, K.C., Chen, D., Robbins, F.M., Wang, L., Larsen, P., Luhm, R., Stroncek, D.F., 2004. Ambiguous allele combinations in HLA Class I and Class II sequence-based typing: when precise nucleotide sequencing leads to imprecise allele identification. J. Transl. Med. 2 (1), 30.

alleles. http://hla.alleles.org/alleles/nulls.html.

Bidwell, J.L., Jarrold, E.A., Laundy, G.J., Klouda, P.T., Bradley, B.A., 1986. Molecular genetics of HLA-DR 'BR': allogenotypes of DR1 and DR'BR' are indistinguishable. Tissue Antigens 27, 99–101.

Breslauer, K.J., Frank, R., Blöcker, H., Marky, L.A., 1986. Predicting DNA duplex stability from the base sequence. Proc. Natl. Acad. Sci. U S A 83, 3746–3750.

Bugawan, T.L., Horn, G.T., Long, C.M., Mickelson, E., Hansen, J.A., Ferrara, G.B., Angelini, A., Erlich, H.A., 1988. Analysis of HLA-DP allelic sequence polymorphism using the in vitro enzymatic NDA amplification of DP-A and DP-B loci. J. Immunol. 141 (12), 4024–4030.

Carlsson, B., Wallin, J., Böhme, J., Möller, E., October 1987. HLA-DR-DQ haplotypes defined by restriction fragment analysis. Correlation to serology. Hum. Immunol. 20 (2), 95–113.

Delfino, L., Morabito, A., Longo, A., Ferrara, G.B., 1998. HLA-C high resolution typing: analysis of exons 2 and 3 by sequence based typing and detection of polymorphisms in exons 1-5 by sequence specific primers. Tissue Antigens 52 (3), 251–259.

Dunn, P.J., 2015. Novel approaches and technologies in molecular HLA typing. In: Bugert, P. (Ed.), Molecular Typing of Blood Cell Antigens, Chapter 18, Methods in Molecular Biology, vol. 1310. Springer Science, New York, pp. 213–230.

ebi. https://www.ebi.ac.uk/ipd/imgt/hla/ambig.html.

Elsner, H.-A., Blasczyk, R., 2004. Immunogenetics of HLA null alleles: implications for blood stem cell transplantation. Tissue Antigens 64, 687–695.

Erlich, H., 2012. HLA DNA typing: past, present, and future. Tissue Antigens 80, 1–11.

Erlich, H., Bugawan, T., Begovich, A.B., Scharf, S., Griffith, R., Saiki, R., Higuchi, R., Walsh, P.S., 1991. HLA-DR, DQ and DP typing using PCR amplification and immobilized probes. Eur. J. Immunogenet. 18, 33–55.

Faner, R., Casamitjana, N., Coll, J., Caro, P., Pujol-Borrell, R., Palou, E., Juan, M., 2006. Real-time PCR using fluorescent resonance emission transfer probes for HLA-B typing. Hum. Immunol. 67, 374–385.

fileadmin. https://www.efi-web.org/fileadmin/user_upload/Website_documenten/EFI_Committees/Standards_Committee/2017-10-31_Standards_version_7.pdf.

Gersuk, V.H., Nepom, G.T., 2006. A real-time PCR approach for rapid high resolution subtyping of HLA-DRB1*04. J. Immunol. Methods 317 (1-2), 64–70.

Heather, J.M., Chain, B., 2016. The sequence of sequencers: the history of sequencing DNA. Genomics 107, 1–8.

igdawg. http://igdawg.org/cwd.html.

Kibbe, W.A., July 2007. OligoCalc: an online oligonucleotide properties calculator. Nucleic Acids Res. 35 (Web Server issue), W43–W46.

Long, E.O., Gorski, J., Rollini, P., Wake, C.T., Strubin, M., Rabourdin-Combe, C., Mach, B., 1983. Molecular analysis of the genes for human Class II antigens of the major histocompatibility complex. Hum. Immunol. 8, 113–121.

Lorenz, T.C., 2012. Polymerase chain reaction: basic protocol plus troubleshooting and optimization strategies. J. Vis. Exp. (63), e3998. https://doi.org/10.3791/3998.

Madden, K., Chabot-Richards, D., 2019. HLA testing in the molecular diagnostic laboratory. Virchows Arch. 474, 139–147.

Maeda, M., Uryu, N., Murayama, N., Ishii, H., Ota, M., Tsuji, K., Inoko, H., 1990. A simple and rapid method for HLA-DP genotyping by digestion of PCR amplified DNA with allele-specific restriction endonucleases. Hum. Immunol. 27, 111–121.

Maxam, A.M., Gilbert, W., February 1977. A new method for sequencing DNA. Proc. Natl. Acad. Sci. U S A 74 (2), 560–564.

Mytilineos, J., Scherer, S., Opelz, G., 1990. Comparison of RFLP-DR beta and serological HLA-DR typing in 1500 individuals. Transplantation 50 (5), 870–873.

nomenclature. http://hla.alleles.org/nomenclature/stats.html.

northwestern. http://biotools.nubic.northwestern.edu/OligoCalc.html.

Nunes, E., Heslop, H., Fernandez-Vina, M., Taves, C., Wagenknecht, D.R., Eisenbrey, A.B., Fischer, G., Poulton, K., Wacker, K., Hurley, C.K., Noreen, H., Hurley, C.K., Sacchi, N., 2011. Definitions of histocompatibility typing terms: harmonization of histocompatibility typing terms working group. Hum. Immunol. 72 (12), 1214–1216.

Olerup, H., Zetterquist, H., 1991. HLA-DRBl*01 subtyping by allele-specific PCR amplification: a sensitive, specific and rapid technique. Tissue Antigens 37, 197–204.

Pollard-Knight, D., Simmonds, A.C., Schaap, A.P., Akhavan, H., Brady, M.A., 1990. Analytical biochemistry: nonradioactive DNA detection on southern blots by enzymatically triggered chemiluminescence. Anal. Biochem. 185 (2), 353–358.

Prober, J.M., Trainor, G.L., Dam, R.J., Hobbs, F.W., Robertson, C.W., Zagursky, R.J., Cocuzza, A.J., Jensen, M.A., Baumeister, K., 1987. A system for rapid DNA sequencing with fluorescent chain-terminating dideoxynucleotides. Science 238 (4825), 336–341.

Saiki, R.K., Scharf, S., Faloona, F., Mullis, K.B., Horn, G.T., Erlich, H.A., Arnheim, N., December 20, 1985. Enzymatic amplification of beta-globin genomic sequences and restriction site analysis for diagnosis of sickle cell anemia. Science 230 (4732), 1350–1354.

Saiki, R.K., Gelfand, D.H., Stoffel, S., Scharf, S.J., Higuchi, R., Horn, G.T., Mullis, K.B., Erlich, H.A., January 29, 1988. Primer-directed enzymatic amplification of DNA with a thermostable DNA polymerase. Science 239 (4839), 487–491.

Saiki, R.K., Walsh, P.S., Levensont, C.H., Erlich, H.A., 1989. Genetic analysis of amplified DNA with immobilized sequence-specific oligonucleotide probes. Proc. Natl. Acad. Sci. U S A 86, 6230–6234.

Sanger, F., Coulson, A.R., 1975. A rapid method for determining sequences in DNA by primed synthesis with DNA polymerase. J. Mol. Biol. 94, 441–448.

Sanger, F., Nicklen, S., Coulson, A.R., December 1977. DNA sequencing with chain-terminating inhibitors. Proc. Natl. Acad. Sci. U S A 74 (12), 5463–5467.

Southern, E.M., 1975. Detection of specific sequences among DNA fragments separated by gel electrophoresis. J. Mol. Biol. 98, 503–517.

Zetterquist, H., Olerup, O., 1992. Identification of the HLA-DRB1*04, -DRB1*07 and -DRB1*09 alleles by PCR amplification with sequence-specific primers (PCR-SSP) in 2 hours. Hum. Immunol. 34, 64–74.

CHAPTER 8

Massively parallel (next generation) sequencing

Contents

Bullet points:

1. Next generation sequencing (NGS) is a simultaneous clonal sequencing of hundreds to thousands of single strands of isolated DNA.
2. Clonal sequencing *reduces* HLA (human leukocyte antigen) typing ambiguities by identifying which polymorphisms are on the same chromosome ("setting the phase").
3. NGS is possible due to advances in computation power, radical new sequencing technologies, and existence of extensive databases of human gene sequences, including the HLA sequences database.
4. There are two primary technological families of NGS methodologies: Sequencing by Synthesis and Sequencing by Hybridization and Ligation.
5. Commonalities among NGS methods for HLA typing include initial polymerase chain reaction (PCR) amplification of whole loci or targeted exons using "barcoded" primers, fragmentation of the target DNA to create a library of sequencing templates, ligation of adapter molecules to the selected fragments, and physical isolation of individual fragments for the sequencing chemistries.
6. Unlike NGS for disease risk (such as cancer promoter mutations), there is no consensus "wild-type" gene sequence to which sequences are compared. HLA typing by NGS must compare sequences obtained against *all* confirmed sequences for the target locus in the international HLA sequence database (European Bioinformatics Institute, https://www.ebi.ac.uk/ipd/imgt/hla/). Sequencing must also exclude pseudogenes and unintended amplifications of shared conserved sequences in other major histocompatibility complex (MHC) loci.

HLA from Benchtop to Bedside
ISBN 978-0-12-823976-6
https://doi.org/10.1016/B978-0-12-823976-6.00008-1

7. HLA NGS typing requires targeted amplification due to the sequence homologies among loci (including noncoding loci) and the extraordinary polymorphism of the HLA genes. (Note Bullet point 6 above.)

Next generation sequencing (NGS) is massively parallel sequencing with hundreds to thousands of simultaneous sequencing reactions. Unlike Sanger sequencing, NGS is single strand sequencing which *should* eliminate ambiguities (see previous discussion in Chapter 6). Most NGS systems rely on redundant sequencing reactions (sequencing overlapping regions of the target or, potentially, genomic DNA) and should reduce misreads which are a result of the inherent error rate of DNA polymerases or ligases used for the sequencing reactions, and also improve the accuracy of the nucleotide assignments. Finally, by selecting the targets used, NGS allows for the potential sequencing of entire genomes. For human leukocyte antigen (HLA) typing, we are only interested in the major histocompatibility complex (MHC) portion of the short arm of chromosome six.

There are common steps in the current NGS methods. The first step is isolation of DNA as is currently done for molecular HLA typing. Then, the isolated DNA is converted to a "library" of sequencing templates by fragmentation of the DNA using shear forces or endonucleases followed by selection of fragments of appropriate size for the sequencing reactions used (as few as 50 and as many as 3000 base pairs) and performing the chemical reaction to ligate an adapter molecule to each of the fragments.

NGS is "clonal sequencing" because one fragment is sequenced in each individual reaction well or spot. Clonality is usually obtained by limiting dilution of the fragment library.

The sequencing templates are spatially separated by immobilization to a solid surface or other method which allows for isolation of the individual reactions, and, finally, there is massive data analysis to allow for alignment of the sequences of thousands of short sequences from each of the thousands of sequencing reactions to provide the final genome sequence. The ability to identify overlapping sequences and align the sequences in massively parallel sequencing is (not simply) due to the power of modern computing and the existence of sequence library databases to which the sequences generated by each method are compared. The generated sequences are evaluated for the quality of the data which is frequently described as the "depth" of the reads at each base pair. Depth is a measure of the number of sequencing reads which returned the same sequence. Another measure is the Phred score (a measure of the quality of the identification of the nucleobases generated by automated DNA sequencing) that is inversely related to the $\log_{10}$ of the likelihood of an incorrect base call. Therefore, the higher the Phred score, the "better" the call or the lower the likelihood that the base call is incorrect (Gandhi et al., 2017).

Some consistent differences between Sanger sequencing and NGS include an increased error rate with A-T rich sequences in NGS (Sanger sequencing errors increase in G-C rich sequences) and increased error rates in NGS with nucleotide repeats (that is, homopolymers) compared to Sanger sequencing.

Most NGS systems incorporate a polymerase chain reaction (PCR) amplification step after immobilization or fluid-phase isolation of the fragments to increase the available signal for detection. As with any PCR step, errors may be introduced during amplification, and some sequences show preferential amplification. Amplifications in NGS for HLA typing may be directed at highly polymorphic exons, whole loci, or span regions of the MHC. Additional common procedural steps include DNA "labels" or tags on the isolated and/or amplified single-stranded DNA that are used for "assembling" the sequences. Assembly utilizes the combination of the computing power available and the databases to which the sequences are compared, and, importantly, the labels on the sequencing products may be used to identify the individual or family of primers used in the initial PCR amplification.

Although there are multiple variations of NGS methods, there are two main NGS methodology families: **Sequencing by Synthesis** and **Sequencing by Hybridization and Ligation** (SHL).

Sequencing by synthesis has three main methodologies, fluorescently labeled, reversible nucleotide terminator chemistry, ion semiconductor sequencing, and single molecule real-time sequencing (SMRT). An early method, pyrosequencing, is no longer commercially available due to successful competition from newer methodologies.

Pyrosequencing (example: Roche 454, (discontinued), Roche, Basel, Switzerland) involves incorporation of a deoxyribonucleotide triphosphate (dNTP) by DNA polymerase which releases pyrophosphate (PP). PP reacts with luciferyl adenylate resulting in an excited state of oxyluciferin in the presence of oxygen. Oxyluciferin returns to its ground state by releasing a photon which is the detection step. This method uses sequential additions of nucleotides and had limited read lengths (<800 bases). Technological advances included isolation of single-stranded DNA on capture beads and amplification before deposition in "picotiter" wells for the sequencing reaction (Moonsamy et al., 2013). The "454" system and chip system was capable of tagging ("barcoding") 48 samples for sequencing and producing up to 2304 PCR products for sequencing in a single run. Amplification was a "4-primer" approach in which target sequences are amplified with one pair of "inner" primers with a common sequence tag on the 5′ primer and a second "outer" primer pair using the same tagged 3′ primer and a "barcoded" 5′ primer (multiplex identifier or MID) with the adapter sequences. The PCR products are pooled, and sequencing is performed on the 4-primer products. Pyrosequencing was the first of the NGS methods commercially introduced and was successful until eclipsed by methods that gave faster, longer, and more reliable base reads.

Fluorescently labeled, reversible nucleotide terminator chemistry uses dye-tagged dNTPs added to the template DNA in the presence of DNA polymerase (example: Illumina MiSeq, Illumina, Inc, San Diego, CA) (Heather and Chain, 2016). The target DNA is fragmented and adapter bracketed prior to being passed over a "lawn" of complementary oligonucleotides which bind individual fragments for solid phase PCR amplification. The solid phase amplification produces clusters of clonal populations in what is called "bridge amplification" because the strands arch over to prime the next round of polymerization with neighboring surface-bound DNA. The sequence of nucleotide incorporation is detected by charge-coupled device (CCD) to determine which fluorophore is at the 3′ hydroxyl position until cleavage of the fluorophore and the next cycle of polymerase and dNTP addition. The polymerase is allowed to incorporate one nucleotide, then the reaction site is washed. Laser excitation is used to detect *which* dye is incorporated at each site, then the terminator and label are chemically removed and the cycle repeats.

Ion Semiconductor Sequencing (example: Ion Torrent, Thermo Fisher Scientific, Waltham, MA) relies on detection of pH change when nucleotides are incorporated. The target oligonucleotides are isolated by limiting dilution, and microwells are set up to contain single molecules of the single-strand DNA. The wells are "flooded" with one of the dNTPs, and DNA polymerase is added. If the corresponding nucleotide is incorporated, a hydrogen ion is released and detected by an ion-selective field-effect transistor in the well (an ultraminiature pH meter). The amount of hydrogen ion released is directly proportional to the amount of nucleotide incorporated. The wells are washed, the next dNTP is added, and the cycle repeated.

Single Molecule Real-time Sequencing uses nanophotonic structures called zero-mode waveguides to allow visualization of single molecule incorporation of base-linked fluorescent nucleotides by template-bound DNA polymerase on isolated target DNA (Eid et al., 2009). Commercial application of the SMRT methodology (Pacific BioSciences of California, Inc, Menlo Park, CA) has been applied to HLA typing using PCR amplification of full-length HLA Class I genes and extended amplifications of HLA Class II genes (Turner et al., 2018). The sequencing library is produced with primers that are DNA bar-coded with attached sequences to allow identification, and the amplicon products are "cleaned up" to remove free nucleotides, primer-dimers, and residual primers. The pooled amplicons are subjected to DNA damage repair, and sequencing adapters (single-strand DNA hairpins) are ligated for orientation of the amplicon and annealing of the sequencing primers. The DNA hairpin adapters result in a circular DNA molecule! The actual sequencing under the zero-mode waveguides takes place on individual magnetic beads to which the isolated circular, single-strand DNA is loaded (MagBeads, Pacific BioSciences). Mayor et al. (2015) report reads greater than 3000 base pairs for amplicons of HLA-A, -B, and −C after 5′ to 3′UTR amplifications. Mayor et al. also reported very high accuracy in reading homopolymers in the amplified sequences.

NGS sequencing by Hybridization and Ligation

This method is included for completeness. An extensive search of the HLA typing literature did not find references indicating the use of this method for HLA typing.

In contrast to sequencing by synthesis, SHL used fluorescently labeled eight base hybridization probes (originated by Agencourt Personal Genomics acquired by Applied Biosystems, and finally acquired and discontinued by Thermo Fisher Scientific, Waltham, MA). Each probe contained two of 16 possible dinucleotide pairs, three modified ("degenerate") bases, and three nonnatural hydrophobic ("universal") bases (the degenerate and hydrophobic bases are always the same). All 16 probes were incubated with template fragments ligated through an "adapter" sequence to a solid phase support (e.g., beads). The adapter primer allowed ligation by a DNA ligase (instead of DNA polymerase) of the first probe to the template when the probe pair matches the DNA sequence of the target. Binding was detected by excitation of the ligated probe label fluorescence. Then, a cleavage reaction removed the three "universal" bases and the fluorescent label and hybridization repeated.

The SHL target sequence is "Interrogated" at positions five bases apart. After seven cycles, the probes are removed and a new set of primers are used one base shorter and, thus, closer to the bead to which the target template is ligated. The process is repeated until all bases in the template are identified (that is, each position is "read" twice) by shortening the primer one base each cycle. SHL methods result in high accuracy over long reads because of the specificity imparted by binding of complementary probes.

Selection of molecular HLA typing methods

The most rapid molecular typing methods are the qPCR and PCR-SSP (Chapter 7). Both methods are suitable for single patient or donor typing and rapidly provide antigen level or low-to-medium resolution typing which is appropriate for deceased donor typing. When higher resolution typing is desired, routine medium to high resolution typing can be obtained with PCR-rSSO (Chapter 7) and is most efficient as batch testing. Commercial rSSO kits have been extended to include additional primer sets to resolve many typing ambiguities that must be clarified to meet accreditation standards. rSSO can also be supplemented with PCR-SSP results to resolve many common typing ambiguities.

When high resolution typing is required, e.g., hematopoietic stem cell transplantation, Sanger sequencing (Chapter 7), or NGS become the most appropriate methods. Ambiguities that prevent allele assignments in Sanger sequencing have been approached using combinations of methods, including PCR-SSP targeting specific single nucleotide polymorphisms and combining the results of rSSO typing with the sequencing results to exclude allele combinations. Ambiguities are much less likely with the clonal typing of NGS and the ability to determine the "phase" of single nucleotide polymorphisms that would result in ambiguities in other molecular typing methods. NGS HLA typing is likely to replace Sanger sequencing as it becomes less expensive to perform. In addition, and significantly, the very low error rates found with NGS HLA typings may further reduce costs because of less need for repeat testing (e.g., confirmatory testing) (Baier et al., 2019).

As a consultant, I would recommend qPCR (rtPCR) for deceased donor typing or HLA allele screening, rSSO for batch routine HLA typing, and adoption of NGS for high resolution typing, particularly if supporting a hematopoietic stem cell transplantation program. As the instrumentation and specimen processing for NGS improve (and cost-per sample decline), laboratories may find that their HLA typing needs are completely covered by qPCR and NGS.

References

Baier, D.M., Hofmann, J.A., Fischer, H., Rall, G., Stolze, J., Ruhner, K., Lange, V., Sauter, J., Schmidt, A.H., 2019. Very low error rates of NGS-based HLA typing at stem cell donor recruitment question the need for a standard confirmatory typing step before donor work-up. Bone Marrow Transplant. 54, 928–930.

Eid, J., Fehr, A., Gray, J., Luong, K., Lyle, J., Otto, G., Peluso, P., Rank, D., Baybayan, P., Bettman, B., Bibillo, A., Bjornson, K., Chaudhuri, B., Christians, F., Cicero, R., Clark, S., Dalal, R., deWinter, A., Dixon, J., Foquet, M., Gaertner, A., Hardenbol, P., Heiner, C., Hester, K., Holden, D., Kearns, G., Kong, X., Kuse, R., Lacroix, Y., Lin, S., Lundquist, P., Ma, C., Marks, P., Maxham, M., Murphy, D., Park, I., Pham, T., Phillips, M., Roy, J., Sebra, R., Shen, G., Sorenson, J., Tomaney, A., Travers, K., Trulson, M., Vieceli, J., Wegener, J., Wu, D., Yang, A., Zaccarin, D., Zhao, P., Zhong, F., Korlach, J., Turner, S., 2009. Real-time DNA sequencing from single polymerase molecules. Science 323 (5910), 133–138.

European Bioinformatics Institute. https://www.ebi.ac.uk/ipd/imgt/hla/.

Gandhi, M.J., Ferriola, D., Huang, Y., Duke, J.L., Monos, D., 2017. Targeted next-generation sequencing for human leukocyte antigen typing in a clinical laboratory. Arch. Pathol. Lab Med. 141 (6), 806–812.

Heather, J.M., Chain, B., 2016. The sequence of sequencers: the history of sequencing DNA. Genomics 107, 1–8.

Mayor, N.P., Robinson, J., McWhinnie, A.J.M., Ranade, S., Eng, K., Midwinter, W., Bultitude, W.P., Chin, C.-S., Bowman, B., Marks, P., Braund, H., Madrigal, J.A., Latham, K., Marsh, S.G.E., 2015. HLA typing for the next generation. PLoS One 10, e0127153.

Moonsamy, P.V., Williams, T., Bonella, P., Holcomb, C.L., Höglund, B.N., Hillman, G., Goodridge, D., Turenchalk, G.S., Blake, L.A., Daigle, D.A., Simen, B.B., Hamilton, A., May, A.P., Erlich, H.A., 2013. High throughput HLA genotyping using 454 sequencing and the Fluidigm Access Array™ system for simplified amplicon library preparation. Tissue Antigens 81 (3), 141–149.

Turner, T.R., Hayhurst, J.D., Hayward, D.R., Bultitude, W.P., Barker, D.J., Robinson, J., Madrigal, J.A., Mayor, N.P., Marsh, S.G.E., 2018. Single molecule real-time DNA sequencing of HLA genes at ultra-high resolution from 126 international HLA and immunogenetics workshop cell lines. HLA 91, 88–101.

CHAPTER 9

HLA alleles, disease associations, and HLA pharmacogenetics

Contents

Bullet points:

1. Correlation does not mean causation; disease associations with genetic markers almost always start as statistical associations without defined mechanisms.
2. Human leukocyte antigen (HLA) alleles have been associated with disease risks and risks of acute drug reactions.
3. There are multiple mechanisms identified and proposed (hypothetical) to explain the role of HLA alleles in disease risk and acute drug reactions. A disease or drug reaction risk association may be due to:
   - **a.** Linkage dissociation with the actual gene causing the risk.
   - **b.** HLA allele restricted abnormal oligopeptide processing or presentation resulting in presentation of normal self-antigens to reactive T cells.
   - **c.** HLA allele restricted presentation of oligopeptides from infectious organisms or viruses that evoke T cell responses to epitopes shared by self-proteins.
   - **d.** Disease-associated modification of self-antigens resulting in presentation of the oligopeptides by those HLA-restricted alleles capable of presenting the abnormal oligopeptides.
   - **e.** Drug or drug metabolite binding in susceptible HLA molecule oligopeptide binding groove resulting in either abnormal presentation of normal oligopeptides or the presentation of normal oligopeptides not normally presented.
4. Linkage dissociation is the nonrandom occurrence of a gene or allele with another gene, allele, risk of or resistance to a disease, or risk of an adverse drug reaction to specific medications.

HLA from Benchtop to Bedside
ISBN 978-0-12-823976-6
https://doi.org/10.1016/B978-0-12-823976-6.00009-3

5. Relative risk and odds ratio are measures of the association of a specific characteristic (e.g., HLA allele) and the risk of developing a disease or adverse drug reaction.

Almost every medical student learns the risk association between HLA-B27 (human leukocyte antigen) positivity and ankylosing spondylitis. How about HLA risk associations with Hodgkin's Lymphoma (HL) and resistance to infection with HIV? Some HLA alleles increase risk of disease, some decrease risk, and some are associated with serious reactions to commonly prescribed medications. Examples of various HLA associations with diseases and pharmacogenetics are discussed in this chapter.

The first disease associated with HLA was HL in French patients (Amiel, 1967) using the limited panel of serological typing reagents available at the time. The HLA first associated with increased risk of classical HL was the leukocyte antigen 4c of Dr. Rose Payne which was subsequently shown to be a broad serological specificity in the HLA-B family of antigens (Thorsby, 2009). It is important to note that Amiel did not claim that being 4c positive was the actual risk factor and that the results should be interpreted "with caution." Subsequent disease association studies have led to current understanding of the complex HLA interactions with HL subtypes and Epstein—Barr virus (EBV) infection status. HLA-A*01 is most consistently associated with increased risk of EBV positive classical HL. HLA-B*37 is the most frequently associated HLA-B locus allelic group (McAulay and Jarrett, 2015). The introduction of molecular typing for HLA and recent addition of genome-wide association studies (GWAS) have further refined the associations of HLA alleles with specific lymphomas, EBV infection status, and population differences in risk (Howell, 2014).

The mathematics of disease associations

The history of the association of increased or decreased risk of HL subtypes with specific HLA alleles (reviewed by McAulay and Jarrett, 2015) is instructive because it demonstrates the importance of understanding linkage disequilibrium (LD), relative risk (RR), and selection bias in population studies. HLA disease associations are affected by the population studied (allele frequencies vary significantly between populations), methods used in HLA typing, and proper selection of case controls.

Linkage disequilibrium is caused by nonrandom association of genes in a population which may be caused by selection pressure on individuals who carry particular alleles of two or more genes or by physical proximity of two loci. Early evaluations of LD assumed there was random mating in the population being studied, the sample population was representative of the population from which it was derived, and there was no need to determine the "phase" of the alleles in the two (or more) loci being studied (Rogers and Huff, 2009). The major histocompatibility complex (MHC) is relatively compact, the physical distances between many of the HLA loci are relatively small, and the HLA genes are usually inherited as haplotypes ("in phase"), with the common exception of the HLA-DP loci (see Chapter 2). Before next generation sequencing (NGS) and the ability to sequence single strands of DNA to "assemble" the nucleotide sequences of chromosomal regions or whole chromosomes, it was exceedingly difficult to determine within a locus and between loci to which chromosome ("phase" or haplotype) an allele belonged. Large, validated databases of HLA haplotype frequencies can be used to *predict* MHC allele associations but cannot replace construction of a pedigree for elucidation of the haplotypes of an individual. With family HLA typing, a pedigree can be readily constructed, and, with personal information removed, incorporated into haplotype databases such as the Allele Frequencies Net Database (http://www.allelefrequencies.net/hla6003a.asp; Gonzalez-Galarza et al., 2015).

The importance of LD should not be underestimated. Hemochromatosis was strongly linked to HLA-A3 in early studies (and used as a risk marker in my early HLA laboratory practice). Fortunately, an allele of the HFE locus, very close to the HLA-A locus, was discovered. The *HFE* gene produces an MHC Class I—like cell surface protein for which single base pair substitutions at C282Y or H63D result in hemochromatosis in homozygous carriers (Porto et al., 1998). In northern Europeans, the variant allele of *HFE*(C282Y) is in very strong LD with HLA-A3.

As noted above, it is usually assumed that alleles sort randomly in the randomly mating population(s), and this would predict zero association between alleles. Nonrandom associations can be tested statistically by comparing frequencies obtained against probability tables. The test of LD I learned in population genetics class is the Lewontin and Kojima (1960), application of the Fisher theorem of allele distributions in populations.

$$D = pr(1-p)(1-r) - p(1-r)r(1-p)$$

D is the measure of LD. D is zero when the genes are not linked. p and r are the frequencies of alleles a and b of the allelic pairs Aa and Bb which are assumed to be at independent loci. Rogers and Huff (2009) discuss alternative formulas and suggest a formula which does not assume knowledge of phase and examine allelic covariance (as a measure of LD) across diploid (e.g., human) genotypes (and assuming no frequency differences between males and females as in HLA alleles).

Relative Risk is a measure of the likelihood that an individual with a genetic marker (such as a specific HLA allele) will get a disease compared to a randomly chosen person from the same population without that allele. Disease association studies may incorrectly associate an HLA allele with disease risk when the actual gene is distinct but physically close (example: hemochromatosis was initially "associated" with HLA-A3 but was actually caused by an allele of the physically close HFE gene) (Porto et al., 1998).

$$RR = \frac{a/(a+b)}{c/(c+d)}$$

$$OR = \frac{a/b}{c/d}$$

RR (relative risk or R_p)

OR (odds ratio)

a is the number of diseased individuals with the allele

b is the number of diseased individuals without the allele

c is the number of controls with the allele

d is the number of controls without the allele (Rimm et al., 1980)

Odds Ratio and RR are not interchangeable. For HLA: *RR* is a measure of frequency of disease when an HLA allele is present compared to controls; *OR* is a measure of the odds of having an allele in those with the disease compared to the odds of the allele being present in the control population. Odds ratios tend to overestimate association between genetic markers (e.g., HLA alleles) and disease risks (Ranganathan et al., 2015).

Statistical significance calculation for *RR* (and *OR*) is from the chi-square distribution. Variance (V) is calculated as $V = 1/a + 1/b + 1/c + 1/d$ and the chi-square calculation is $(1/V) \times (\log_e RR)^2$ (usually at one degree of freedom) (Dyer and Warrens, 1994). For small values of *a*, *b*, *c*, or *d*, the Haldane correction can/should be applied (Haldane, 1956).

$$RR = \frac{(2a+1)(2d+1)}{(2b+1)(2c+1)} \quad \text{Variance } (V) = \frac{1}{a+1} + \frac{1}{b+1} + \frac{1}{c+1} + \frac{1}{d+1}$$

Table 2 Chi-square table from Amiel, J.L., (1967).

Frequency of 4c antigen in patients with Hodgkin's disease and in normal subjects		
	Positive	**Negative**
Patients with Hodgkin's disease	21	20
Normal subjects	12	32

The original report of the association of risk of HL with an HLA used simple chi-square (X^2) calculation (Table 2) on 41 patients with HL and 44 "nonrelated" normal subjects to calculate the significance as $X^2 = 5.06$ at one degree of freedom and $P < .05$. Applying the Haldane correction (above) gives RR = 8.16 and $P < .05$.

How do HLA alleles cause disease?

As noted many times before in this book and in almost every paper and publication about the human MHC and the HLA genes, this is the most polymorphic system in the human genome. It should not be surprising that there are multiple hypothetical mechanisms for diseases and drug sensitivities to be causally associated with HLA. The proposed mechanisms can be grouped as:

- Linkage disequilibria with the gene(s) that actually cause the disease (example: hemochromatosis (HFE gene) which was first associated with HLA-A3) (Porto et al., 1998).
- Presentation of self-antigens to nondeleted self-reactive T cell clones (example: pancreatic peptides not presented in the thymus in Type 1 Diabetes Mellitus) (Wang et al., 2017).
- Presentation of bacterial or viral oligopeptides that mimic self-peptides (example: HLA-B27 and Ankylosing spondylitis) (Oldstone, 1998).
- Presentation of damaged (altered) self-oligopeptides (example: citrullinated chromogranin peptides in Type 1 Diabetes Mellitus (Yin et al., 2013).
- Shared epitope (SE) regions with conserved "cusp-like" structure serve as ligands for T cell interactions permitting abnormal oligopeptide presentation (example: Rheumatoid arthritis (RA)) (Holoshitz, 2013).
- Drug binding to HLA molecules and interfering with oligopeptide binding and presentation (example: Abacavir adverse drug reactions in patients positive for HLA-B*57:01) (Ramsbottom et al., 2018).

- Inability of either the HLA allele to present the relevant oligopeptide or the T cell receptor to recognize the complex.

HLA allele disease associations are not all or nothing. Although odds ratios and RRs may be very high for associations of an allele with a disease, it is not uncommon for individuals with the risk-associated HLA allele(s) to not have evidence of the disease. In addition, the risks for disease and the alleles associated with the disease risk may vary significantly between populations, highlighting the critical importance of properly selecting controls for statistical risk assessments. The association of HLA-B27 with risk for ankylosing spondylitis varies significantly between different population groups for the many HLA-B*27 alleles tested, even for the most consistently associated HLA-B*27:04 allele (Yang et al., 2014). Alleles of non-MHC genes associated with peptide processing for MHC Class I presentation (e.g., endoplasmic reticulum aminopeptidase, ERAP1) and variants of IL-23 have been associated with increased risk of ankylosing spondylitis and other autoimmune disorders (Robinson and Brown, 2014).

HLA alleles also affect response to cancer treatments such as checkpoint blockade immunotherapies targeting cytotoxic T lymphocyte-associated protein 4 (CTLA-4), programmed cell death protein 1 (PD-1), or its ligand (PDL-1). Homozygosity at HLA-A, -B *or* -C, *or* HLA-DPB is associated with reduced survival in patients with advanced melanoma or nonsmall cell lung cancer. Tumor loss of heterozygosity at HLA Class I loci is associated with decreased survival. Melanoma patients with HLA-B*15:01 had significantly reduced survival, and patients with HLA-B44 (e.g., HLA-B*44:02) superfamily alleles had significantly better survival (Chowell et al., 2018).

Critical readers should be wary of HLA disease associations from older studies, particularly when the risk was associated with HLAs (serological methods) and not reproduced in newer studies using molecular methods (including GWAS). Additionally, small sample sizes and associations in single populations can be very misleading since, as shown in Table 3, with HLA-B*58:01 risk varies significantly between different population groups being sampled. HLA allele associations with diseases can be very different in different populations. The review by Matzaraki et al. (2017) is an excellent starting point with a longer list of disease associations and the online HLA-drug adverse reaction database http://www.allelefrequencies.net/hla-adr/default.asp is continually updated (Ghattaoraya et al., 2016).

Table 3 Representative HLA and disease associations. (Note: the table is not comprehensive and is selected for associations with published odds ratios.)

HLA allele	Disease	Odds ratio (OR)	References
HLA-A*01:01	EBV positive classical Hodgkin's Lymphoma	2.33	Huang et al. (2012)
HLA-A24	Thymoma	9.70	Machens et al. (1999)
HLA-A*29:02	Birdshot retinopathy	157.50	Herbort et al. (2017), Levinson et al. (2004)
HLA-A*31:01	Carbamazepine (CBZ)-induced severe cutaneous adverse reactions (SCARS)	Japanese: 33.9; Europeans: 25.93	Kaniwa and Saito (2013)
HLA-B*15:02	Carbamazepine (CBZ)-induced severe cutaneous adverse reactions (SCARS)	Han Chinese: 2504	Kaniwa and Saito (2013)
HLA-B27 (*not* HLA-B*27:06 or B*27:09)	Ankylosing spondylitis	Heterozygous 50, homozygous 100	Brown et al. (2016)
HLA-B51	Behçet's disease	5.78	De Menthon et al. (1999)
HLA-B*57:01	Abacavir hypersensitivity	117.00	Mallal et al. (2002)
HLA-B*58:01	Allopurinol-induced severe cutaneous adverse reactions (SCARS)	Han Chinese: 580.3; Europeans: 80	Kaniwa and Saito (2013)
HLA-DRB1*15:01 (*15:03 in African Americans)	Multiple sclerosis	1.88—2.45	Chi et al. (2019)
HLA-DQB1*06:02	Narcolepsy	26.00	Capattini et al. (2018), Siebold et al. (2004)
HLA-DQB1*05:02	Myasthenia gravis	5.45	Testi et al. (2012)

References

Allele Frequencies Net Database. http://www.allelefrequencies.net/hla6003a.asp.

Amiel, J.L., 1967. Study of the leukocyte phenotypes in Hodgkin's disease. In: Terasaki, P. (Ed.), Histocompatibility Testing, 1967. Munksgaard, Cophenhagen, pp. 79–81.

Brown, M.A., Kenna, T., Wordsworth, B.P., 2016. Genetics of ankylosing spondylitis—insights into pathogenesis. London Nat. Rev. Rheumatol. 12 (2), 81–91.

Capittini, C., De Silvestri, A., Terzaghi, M., Scotti, V., Rebuffi, C., Pasi, A., Manni, R., Martinetti, M., Tinelli, C., 2018. Correlation between HLA-DQB1*06:02 and narcolepsy with and without cataplexy: approving a safe and sensitive genetic test in four major ethnic groups. A systematic meta-analysis. Sleep Med. 52, 150–157.

Chi, C., Shao, X., Rhead, B., Gonzales, E., Smith, J.B., Xiang, A.H., Graves, J., Waldman, A., Lotze, T., Schreiner, T., Weinstock-Guttman, B., Aaen, G., Tillema, J.M., Ness, J., Candee, M., Krupp, L., Gorman, M., Benson, L., Chitnis, T., Mar, S., Belman, A., Casper, T.C., Rose, J., Moodley, M., Rensel, M., Rodriguez, M., Greenberg, B., Kahn, L., Rubin, J., Schaefer, C., Waubant, E., Langer-Gould, A., Barcellos, L.F., January 17, 2019. Admixture mapping reveals evidence of differential multiple sclerosis risk by genetic ancestry. PLoS Genet. 15 (1), e1007808. https://doi.org/10.1371/journal.pgen.1007808. eCollection 2019 Jan.

Chowell, D., Morris, L.G.T., Grigg, C.M., Weber, J.K., Samstein, R.M., Makarov, V., Kuo, F., Kendall, S.M., Requena, D., Riaz, N., Greenbaum, B., Carroll, J., Garon, E., Hyman, D.M., Zehir, A., Solit, D., Berger, M., Zhou, R., Rizvi, N.A., Chan, T.A., 2018. Patient HLA class I genotype influences cancer response to checkpoint blockade immunotherapy. Science 359 (6375), 582–587.

De Menthon, M., Lavalley, M.P., Maldini, C., Guillevin, L., Mahr, A., 2009. HLA–B51/B5 and the risk of Behçet's disease: a systematic review and meta-analysis of case–control genetic association studies. Arthritis Rheum. 15 (10), 61. https://doi.org/10.1002/art.24642.

Dyer, P., Warrens, A., 1994. Design and interpretation of studies of the major histocompatibility complex in disease. In: Lechler, R. (Ed.), HLA and Disease. Academic Press, Harcourt Brace & Company, London, pp. 93–121 (Chapter 6).

Ghattaoraya, G.S., Dundar, Y., González-Galarza, F.F., Maia, M.H., Santos, E.J., da Silva, A.L., McCabe, A., Middleton, D., Alfirevic, A., Dickson, R., Jones, A.R., 2016. A Web Resource for Mining HLA Associations with Adverse Drug Reactions: HLA-ADR. Database (Oxford). http://www.allelefrequencies.net/hla-adr/default.asp. (Accessed 2 April 2019).

Gonzalez-Galarza, F.F., Takeshita, L.Y., Santos, E.J., Kempson, F., Maia, M.H., da Silva, A.L.S., e Silva, A.L.T., Ghattaoraya, G.S., Alfirevic, A., Jones, A.R., Middleton, D., 2015. Allele frequency net 2015 update: new features for HLA epitopes, KIR and disease and HLA adverse drug reaction associations. Nucleic Acids Res. 43 (D1), D784–D788. https://doi.org/10.1093/nar/gku1166.

Haldane, J.B.S., 1956. The estimation and significance of the logarithm of a ratio of frequencies. Ann. Hum. Genet. 20 (4), 309–311.

Herbort Jr., C.P., Pavésio, C., LeHoang, P., Bodaghi, B., Fardeau, C., Kestelyn, P., Neri, P., Papadia, M., July 2017. Why birdshot retinochoroiditis should rather be called 'HLA-A29 uveitis'? Br. J. Ophthalmol. 101 (7), 851–855. https://doi.org/10.1136/bjophthalmol-2016-309764. Epub 2017 Mar 17.

Holoshitz, J., 2013. The quest for better understanding of HLA-disease association: scenes from a road less traveled by. Discov. Med. 16 (87), 93–101.

Huang, X., Kushekhar, K., Nolte, I., Kooistra, W., Visser, L., Bouwman, I., Kouprie, N., Veenstra, R., van Imhoff, G., Olver, B., Houlston, R.S., Poppema, S., Diepstra, A., Hepkema, B., van den Berg, A., 2012. HLA associations in classical Hodgkin lymphoma: EBV status matters. PloS One 7 (7), e39986. https://doi.org/10.1371/journal.pone.0039986.

Howell, W.M., 2014. HLA and disease: guilt by association. Int. J. Immunogenet. 41, 1–12.

Kaniwa, N., Saito, Y., 2013. Pharmacogenomics of severe cutaneous adverse reactions and drug-induced liver injury. J. Hum. Genet. 58, 317–326.

Levinson, R.D., Rjalingam, R., Park, M.S., Reed, E.F., Gjertson, D.W., Kappel, P.J., See, R.F., Rao, N.A., Holland, G.N., 2004. Human leukocyte antigen A29 subtypes associated with birdshot retinochoroidopathy. Am. J. Ophthalmol. 138, 631–634.

Lewontin, R.C., Kojima, K.-I., 1960. The evolutionary dynamics of complex polymorphisms. Evolution 14 (4), 458–472.

Machens, A., Löliger, C., Pichlmeier, U., Emskötter, T., Busch, C., Izbicki, J.R., 1999. Correlation of thymic pathology with HLA in myasthenia gravis. Clin. Immunol. 91 (3), 296–301.

Mallal, S., Nolan, D., Witt, C., Masel, G., Martin, A.M., Moore, C., Sayer, D., Castley, A., Mamotte, C., Maxwell, D., James, I., Christiansen, F.T., 2002. Association between presence of HLA-B*5701, HLA-DR7, and HLA-DQ3 and hypersensitivity to HIV-1 reverse-transcriptase inhibitor abacavir. Lancet 359 (9308), 722–732.

Matzaraki, V., Kumar, V., Wijmenga, C., Zhernakova, A., 2017. The MHC locus and genetic susceptibility to autoimmune and infectious diseases. Genome Biol. 18 (1), 76.

McAulay, K.A., Jarrett, R.F., 2015. Human leukocyte antigens and genetic susceptibility to lymphoma. Tissue Antigens 86 (2), 98–113.

Oldstone, M.B.A., 1998. Molecular mimicry and immune-mediated diseases. Faseb. J. 12 (13), 1255–1265.

Porto, G., Alves, H., Rodrigues, P., Cabeda, J.M., Portal, C., Ruivo, A., Justiça, B., Wolff, R., De Sousa, M., 1998. Major histocompatibility complex class I associations in iron overload: evidence for a new link between the HFE H63D mutation, HLA-A29, and non-classical forms of hemochromatosis. Immunogenetics 47, 404–410.

Ramsbottom, K.A., Carr, D.F., Jones, A.R., Rigden, D.J., 2018. Critical assessment of approaches for molecular docking to elucidate associations of HLA alleles with adverse drug reactions. Mol. Immunol. 101, 488–499.

Ranganathan, P., Aggarwal, R., Pramesh, C.S., 2015. Common pitfalls in statistical analysis: odds versus risk. Perspect. Clin. Res. 6 (4), 222–224. https://doi.org/10.4103/2229-3485.167092.

Rimm, A.A., Hartz, A.J., Kalbfleisch, J.H., Anderson, A.J., Hoffmann, R.G., 1980. Basic Biostatistics in Medicine and Epidemiology. Appleton-Century-Crofts, New York, pp. 296–301 (Chapter 19).

Robinson, P.C., Brown, M.A., 2014. Genetics of ankylosing spondylitis. Mol. Immunol. 57, 2–11.

Rogers, A.R., Huff, C., 2009. Linkage disequilibrium between loci with unknown phase. Genetics 182, 839–844.

Siebold, C., Hansen, B.E., Wyer, J.R., Harlos, K., Esnouf, R.E., Svejgaard, A., Bell, J.I., Strominger, J.L., Jones, E.Y., Fugger, L., 2004. Crystal structure of HLA-DQ0602 that protects against type 1 diabetes and confers strong susceptibility to narcolepsy. Proc. Natl. Acad. Sci. USA 101 (7), 1999–2004.

Testi, M., Terracciano, C., Guagnano, A., Testa, G., Marfia, G.A., Pompeo, E., Andreani, M., Massa, R., 2012. Association of HLA-DQB1*05:02 and DRB1*16 alleles with late-onset, nonthymomatous, AChR-Ab-positive myasthenia gravis. Autoimmune Dis. 2012, 541760. https://doi.org/10.1155/2012/541760. Epub 2012 Oct 9.

Thorsby, E., 2009. A short history of HLA. Tissue Antigens 74, 101–116.
Wang, Z., Xie, Z., Lu, Q., Chang, C., Zhou, Z., 2017. Beyond genetics: what causes type 1 diabetes. Clin. Rev. Allergy Immunol. 52, 273–286.
Yang, T., Duan, Z., Wu, S., Liu, S., Zeng, Z., Li, G., Wang, S., Fan, D., Ye, D., Xu, S., Zhang, L., Pan, F., 2014. Association of HLA-B27 genetic polymorphisms with ankylosing spondylitis susceptibility worldwide: a meta-analysis. Mod. Rheumatol. 24 (1), 150–161.
Yin, L., Dai, S., Clayton, D., Gao, W., Wang, Y., Kappler, J., Marrack, P., 2013. Recognition of self and altered self by T cells in autoimmunity and allergy. Protein Cell 4 (1), 8–16.

CHAPTER 10

Engraftment monitoring

Contents

Bullet points:

1. Engraftment monitoring is used to detect and quantify chimerism after hematopoietic stem cell transplantation.
2. The most commonly used clinical method for engraftment monitoring utilizes polymorphic genetic markers called short tandem repeats (STR) which are PCR amplified, and the fluorescently labeled products are separated and quantified in capillary electrophoresis.
3. Engraftment analysis may be performed on DNA extracted from isolated lineage-specific cell fractions or whole nuclear DNA isolated from bone marrow or peripheral blood samples.
4. Commercial STR kits are designed and optimized for forensic identity testing, not clinical engraftment monitoring.
5. Each laboratory must determine optimal conditions for testing, including optimal DNA concentration for the multiplex amplifications and the minimum percentage of a mixture that may be reported (test sensitivity).
6. If possible, three informative markers should be used to determine the percent chimerism (required by some accrediting agencies). The coefficient of variation (CV) should be less than or equal to 5% for the three selected STR markers.
7. Common testing and biological artifacts in PCR amplification and capillary electrophoresis include stutter products, allelic dropout, and saturation.
8. Alternative methods of chimerism analysis utilizing SNPs, InDels, point mutations, rearrangements, and gene overexpression markers may replace capillary electrophoresis analysis of STR loci in the future.

HLA from Benchtop to Bedside
ISBN 978-0-12-823976-6
https://doi.org/10.1016/B978-0-12-823976-6.00010-X

A little history

The term "chimera" comes from Greek mythology and referred to animals that displayed a mixed phenotype such as a beak and wings with the body, legs, and tail of a lion. In medicine (and modern biology) a chimera is an individual that has two or more genetically distinct cell lineages. Natural chimeras occur from embryonic fusion and survival of fetal cells in the maternal circulation. Iatrogenic chimeras occur from transplantation, especially hematopoietic stem cell transplantation, but may also occur from solid organ transplants and transfusions. Chimerism is the intent of engraftment, and the degree of replacement of the recipient's circulating cell populations is a measure of the success of transplantation.

Engraftment monitoring is a common non-HLA function of histocompatibility laboratories, most likely to be performed in laboratories that support hematopoietic stem cell transplantation. The roots of engraftment monitoring are in identity testing, a forensic procedure, formerly called parentage testing. Modern engraftment monitoring, also called chimerism testing or chimerism analysis, relies on identification of polymorphic genetic markers with alleles that are different and readily distinguishable between the donor(s) and the recipient. Various clinical methods have been validated and described. The most common and extensively validated method uses multiplex polymerase chain reaction (PCR) amplification of well-characterized short tandem repeat (STR) markers. Engraftment monitoring is most clinically useful when performed as a quantitative assessment. Common technical and interpretation challenges are presented with recommendations for clinical use in hematopoietic stem cell transplantation.

Microsatellite DNA, VNTRs, STRs

Microsatellite DNA is common in all organisms that have been evaluated but differ in number, size, and distribution (Chambers and MacAvoy, 2000; Rakoczy-Trojanowska and Bolibok, 2004). Microsatellites are composed of tandemly arranged repetitive DNA sequences (generally four to six nucleotides) up to a kilobyte in length (Fig. 33), most frequently located in untranslated regions (UTRs) and are characterized by being flanked by highly conserved sequences. Microsatellites can be "simple"or "pure:" constructed of repeats of one or more nucleotides; "compound:" constructed of one, two, or more types of repeat units; and "complex:" with variations in repeated unit lengths and interruptions (Chambers and MacAvoy, 2000). The simplest representative formula for microsatellite loci is $(N_1N_2N_3 \ldots N_x)_n$ and the repeat sequence(s) identified alphabetically,

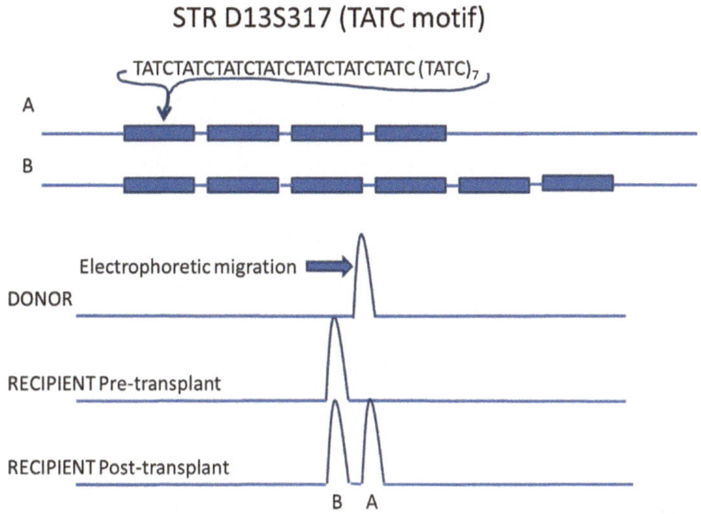

Figure 33 Cartoon of the repetitive structure of microsatellite DNA. A representation of the capillary electrophoresis mobility of the PCR products from amplication is shown below. The structure of an STR can be simple repeats or complex structures of differing lengths and interruptions (see text).

e.g., -$(CA)_n$-. A variation in the classification of STRs and class examples are given on the STRBase website (https://strbase.nist.gov/intro.htm).

For historical reference, "minisatellite DNA" was recognized as a genetic marker by Alec J. Jeffreys and quickly applied to forensic identity testing by the creation of DNA "fingerprints" (Jeffreys et al., 1985a, 1985b) using radiolabeled oligonucleotide probes and restriction endonuclease fractionated DNA separated by electrophoresis in an agarose gel. The introduction of PCR, coupled with expansion of the databases of the human genome (Dib et al., 1996), rapidly expanded the number of microsatellites identified for use in forensic identity testing, population genetics, and clinical medicine. Today, commercial panels of primers for well-documented STRs have become standard of practice for forensic identity testing and widely adopted for engraftment monitoring.

STRs are named for the gene locus with which they are physically associated (e.g., VWA, in the 40th intron of the von Willebrand Factor gene) or the chromosomal location of the locus (e.g., D13S317, "D" for DNA, "13" for chromosome 13, "S" for STR, and "317" for a unique identifier). Note that variants of STRs may be identified by the method used (i.e., CE for capillary electrophoresis) and the measured fragment length. An allele of D13S317 may be named D13S317-CE12 to identify the allele length as measured in capillary electrophoresis (van der Gaag and der Knijff, 2015). STRs may also be identified using massively parallel or "next generation" sequencing (MPS or NGS). There are, however, competing systems for naming STRs which can result in confusion when comparing published information (see: Comment on Nomenclature for STR Alleles and Repeat Structure, https://strbase.nist.gov/intro.htm).

Why STRs instead of VNTRs (variable number of tandem repeats)? Partially, this is a challenge of definitions because the terms have been used interchangeably, although a consensus appears to be that the core number of repeats in VNTRs (minisatellites) is 7–10 nucleotides while the number of repeats is two to six in STRs (microsatellites) (Chambers and MacAvoy, 2000; Rakoczy-Trojanowska and Bolibok, 2004). The STRs in the commercial forensic panels are selected for relatively short PCR product sizes (100–500 base pairs), well-documented alleles at each locus, and compatibility of the primer sets in multiplex amplification (Butler, 2007).

The biological function(s) of microsatellites remain unclear but some locus-specific variants are associated with neurological diseases (e.g., Friedreich's Ataxia and Huntington's Disease), and length variants are associated with abnormal regulation of expression of some genes (e.g., reduced expression of bilirubin UDP-glucuronosyltransferase 1 in Gilbert's Syndrome) (Bagshaw, 2017).

Engraftment monitoring

Engraftment monitoring is clinical laboratory testing, not forensic identity testing. The commercial STR kits (e.g., AmpFLSTR™ Identifiler™, Thermo Fisher Scientific, Waltham, MA, and PowerPlex® variants, Promega Corporation, Madison, WI) provide premixed primers and standardized master mix (polymerase, buffers, and dNTPs), an allelic ladder for product size calibration and include uniform "core" STR markers for forensic identity testing (Butler, 2007; Butler and McCord, 2006). There are no requirements to use commercial kits or primer sets for engraftment monitoring. Laboratory selected and developed primer sets for clinically useful STR markers can be used successfully.

Commercial STR kits are designed for use in forensic identity testing, not for clinical monitoring of patients who have received hematopoietic stem cell transplants. Individual STR loci in the forensic kits may be irrelevant to a given patient due to chromosomal rearrangements or loss of chromosomes in some malignancies and myelodysplastic syndromes (e.g., loss of chromosomes 5 and 7 which would result in D5S818 and D7S820 appearing to indicate only donor lineage is present) (Butler, 2007).

In addition, forensic STR kits are optimized for small amounts of DNA (1–2 ng), corresponding to the nuclear DNA of only 150–300 cells (Clark et al., 2014). Sampling the DNA of small numbers of cells in a patient sample can lead to sample bias and underrepresent a population that is present (either donor or recipient). Larger samples (10–25 ng of DNA) give more reproducible results but may also result in overloading and saturation preventing quantitation of the proportion of donor DNA to recipient DNA (Clark et al., 2014).

Since 2017, the "core" loci in the combined DNA index system (CODIS) of the United States Federal Bureau of Investigation (FBI) has included CSF1PO, FGA, THO1, TPOX, VWA, D3S1358, D5S818, D7S820, D8S1179, D13S317, D16S539, D18S51, D21S11, D1S1656, D2S441, D2S1338, D10S1248, D12S391, D19S433, D22S1045, and amelogenin. European STR standards for forensic testing include pentameric repeat STRs Penta D and Penta E. Commercial STR kits may also include gender markers such as Y-indel (Y chromosome insertions and deletions) and DYS391.

Engraftment monitoring by capillary electrophoresis

As noted previously, the most commonly used methodology for engraftment monitoring uses dye-labeled primer PCR amplification of well-documented and well-characterized STR loci and separation of the amplification products by capillary electrophoresis for identification and quantification. The instrumentation used is the same as used in Sanger sequencing in most clinical laboratories (see Chapter 7). Commercial kits (forensic identity testing) and laboratory-developed primer sets are selected for the conserved regions flanking the STR loci enabling identification of the products by color and size after multiplex amplification.

Samples for testing are directed by the transplanting physicians and may be determined by the patients' underlying disease. Samples may be from bone marrow aspiration or peripheral blood (rarely, buccal swabs or other sources). The samples may be analyzed unfractionated or on samples in which specific cell types (e.g., lymphoid or myeloid fractions) are analyzed after separation. Generally, three samples are analyzed and compared to determine engraftment: patient pretransplant, donor, and patient post-transplant. The pretransplant patient and donor samples are evaluated to identify STR alleles that are different in the patient and donor (Fig. 34).

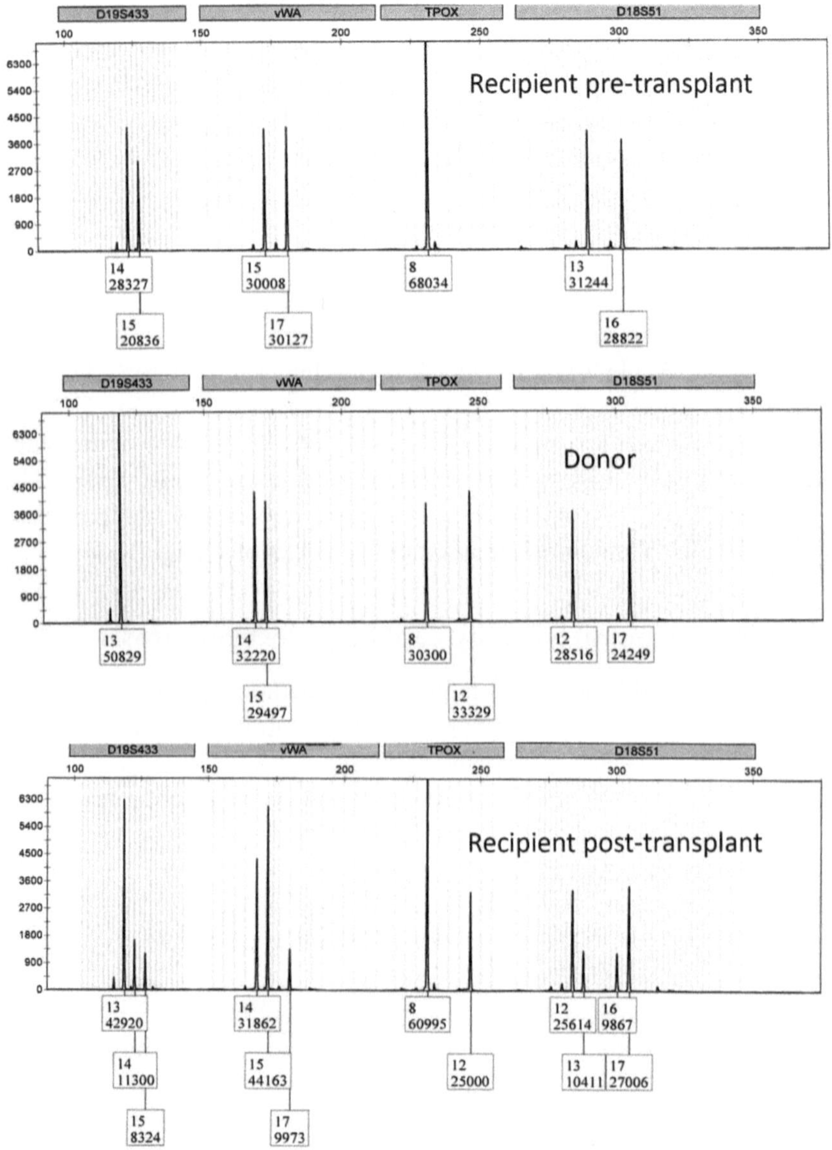

Figure 34 Capillary electrophoresis histograms from STR analysis of hematopoietic transplantation engraftment. Four STR markers are shown. The alleles (top number in boxes under the histograms) are identified by mobility in comparison to a standard mobility "ladder." The areas under the curves are given below the allele number. The short peaks before or after the taller peaks are "stutter" and must be distinguished from alleles (see text). Capillary electrophoresis performed on an Applied Biosystems™ 3500xL model sequencer, GeneMapper™ software, and AmpFLSTR™ Identifiler™ kits for the amplification (Thermo Fisher Scientific, Waltham, MA). Deidentified images used with permission of Sharon Skorupski, BS(CHS).

Analysis of the electrophoresis histogram is greatly simplified by software from many vendors, including the vendors who supply the forensic analysis reagent kits. Identification of informative loci is the responsibility of the laboratorians performing the assay.

The "simplest" interpretive step is in the identification of the loci at which the donor and recipient are different. In Fig. 34, the donor and recipient have entirely different alleles of D19S433 and D18S51. The patient has D19S433-14 and -15, and the donor appears to be homozygous for D19S433-13. The small peak to the left of the D19S433-13 peak is most likely a "stutter" peak caused by slipped strand mispairing (SSM) during amplification (Brookes et al., 2012), but care should be taken to under- or overinterpret artifacts. The recipient also has D18S51-13 and -16 alleles, and the donor has D18S51-12 and -17. In addition to the stutter, the histogram also shows possible saturation at TPOX-8 in the recipient pretransplant sample and D19S433-13 in the donor, both of which appear to be homozygous loci and may cause problems in calculation of the percentage of donor DNA in the recipient's posttransplant analysis. Apparent homozygosity can be the result of preferential amplification or allele dropout due to a mutation in the flanking regions of the locus where the primers are supposed to bind.

In the example (Fig. 34), only D18S51 is "fully informative" in that the recipient and donor have no common alleles and no homozygous alleles (Clark et al., 2014). The percent donor present in the posttransplant sample would be calculated $\frac{(C+D)}{(C+D)+(A+B)} \times 100$ where C and D are the donor alleles and A and B are the recipient's alleles at each locus. The D19S433, VWA, and TPOX loci are "informative" or "semiinformative" because one allele is shared between the donor and recipient at both. The donor and recipient share VWA-15 and TPOX-8. For the VWA locus, the shared allele (VWA-15) is ignored and the percent donor in the posttransplant sample is calculated $\frac{C}{A+C} \times 100$ where C is the donor allele and A is the recipient allele. For the TPOX locus, the recipient is homozygous for TPOX-8 and the donor is heterozygous for TPOX-8 and -12. The percent donor in the posttransplant sample is then calculated $\frac{C}{\left(\frac{A-C}{2}\right)+C} \times 100$ where A is the shared allele (TPOX-8) and C is the unshared allele (TPOX-12) (Clark et al., 2014).

Peak height and area under the curve are proportional. Automated analysis software gives very accurate values for both (excluding saturated peaks and peak heights out of range). Choice of the parameter to use in calculations is, appropriately, a local decision, but should be consistent within an institution.

It is recommended in the UK guidelines that three "fully informative" loci be used in calculation of the percentage of donor DNA in the posttransplant sample and that results with variation of ≤5% between loci be used in the calculations (Clark et al., 2014). Standards of some international laboratory accrediting organizations (e.g., College of American Pathologists) require use of three informative loci. Additional STR loci may need to be analyzed if informative loci are not present in the initial analysis. If clinically necessary, the results of the "semiinformative" loci may be included in the calculations. Because of the apparent saturation of TPOX-8 in the posttransplant sample example (Fig. 34), the TPOX locus should not be used in calculation for this recipient's posttransplant sample.

In the example selected for Fig. 34, TPOX would not be used in the calculations because of saturation, D19S433 would not be used because the result (68%) is more than 5% variant from VWA (76%) and D18S51 (76%). Additional informative loci would need to be analyzed (likely to have been performed using modern STR kits and markers in most laboratories). If the TPOX locus was necessary for calculation, the sample could be rerun on the capillary electrophoresis after dilution.

Alternative platforms for engraftment monitoring

Capillary electrophoresis of fluorescently labeled PCR amplification products from STR loci is well-established for routine engraftment monitoring. For clinical needs, routine STR PCR may lack the sensitivity desired by the physicians caring for the transplant recipient, particularly when there is high risk of relapse. Minimal residual disease (MRD) analysis has driven refinements in STR PCR testing and encouraged the development of potentially more sensitive techniques. The sensitivity of capillary electrophoresis STR PCR is generally cited as between 1% and 5%, though Alizedeh et al. (2002) report 0.4%–5% and Thiede et al. (2001) reported sensitivity below a 10th of a percent by combining STR PCR with fluorescence activated cell sorting to isolate the CD34 positive cell fraction for the STR analysis.

As with HLA typing and many other clinical applications (such as testing for SARS-CoV-2 viral RNA or bacterial DNA), real-time PCR (qPCR or rtPCR, see Chapter 8) can be used to quantify the percentage of donor DNA in posttransplantation samples using biallelic pairs of short insertion/deletion (InDel) polymorphisms instead of the STR loci (Alizadeh et al., 2002). Standard curves are generated, and the cycle threshold (Ct) set for each well-characterized marker pair. Primers for InDel sets are commercially available and the InDels can be selected for both autosomal and sex chromosome markers and linkage independence. The sensitivity of qPCR for chimerism analysis is reported to be 0.1% in the frequently cited paper by Alizadeh et al. (2002).

A more recent addition to the qPCR family is digital droplet PCR (ddPCR) in which the sample DNA is diluted with the amplification master mix, and then microdroplets ("nanodroplets") are physically isolated before amplification (Li et al., 2018). Quantitation is by counting positive wells.The reported advantages of ddPCR include faster assay time than qPCR with equal sensitivity but greater accuracy for detection of relapse (Mika et al., 2019). Like qPCR, ddPCR requires non-STR genetic markers such as InDels for identification of donor and recipient DNA.

MPS or NGS has also been applied to engraftment monitoring and testing. The markers most frequently used for NGS analysis of donor engraftment have been single nucleotide polymorphisms (SNPs) and InDels. The technique has been reported to give sensitive detection of chimersim and is comparable to qPCR results (Askar et al., 2019).

The alternative methods of chimerism analysis are proving as or more sensitive than the more sensitive modifications of capillary electrophoresis PCR of SCT and readily adaptable to highly sensitive MRD analysis. An additional advantage of qPCR, ddPCR, and NGS methods is the ability to analyze disease-associated clonal abnormalities such as InDels, point mutations, rearrangements, and gene overexpression (e.g., NPM1 (Nucleophosmin 1)) (Bader et al., 2019).

References

Alizadeh, M., Bernard, M., Danic, B., Dauriac, C., Birebent, B., Lapart, C., Lamy, T., Le Prise, P.-Y., Beauplet, A., Bories, D., Semana, G., Quelvennec, E., 2002. Quantitative assessment of hematopoietic chimerism after bone marrow transplantation by real-time quantitative polymerase chain reaction. Blood 99 (12), 4618–4625.

Askar, M.Z., Willis, A., Williams, J., Pereira, S., Smith, A., Pyo, C.-W., Nelson, W., Geraghty, D., 2019. An ultrasensitive next generation sequencing method for the detection of minimal residual disease and the assessment of chimerism following hematopoietic cell transplantation. Biol. Blood Marrow Transplant. 25 (3), S325.

Bader, P., Kreyenberg, H., Ossenkoppele, G., 2019. Monitoring minimal residual disease in ALL and AML. In: Carreras, E., Dufour, C., Mohty, M., Kröger, N. (Eds.), The EMBT Handbook, pp. 431–436.

Bagshaw, A.T.M., 2017. Functional mechanisms of microsatellite DNA in eukaryotic genomes. Genome Biol. Evol. 9 (9), 2428–2443.

Brookes, C., Bright, J.-A., Harbison, S.A., Buckleton, J., 2012. Characterising stutter in forensic STR multiplexes. Forensic Sci. Int. Genet. 6, 58–63.

Butler, J.M., 2007. Short tandem repeat typing technologies used in human identity testing. Biotechniques 43 (4), Sii–Sv.

Butler, J.M., McCord, B.R., February 20, 2006. STR Biology, Markers, and Methods. AAFS 2006 Workshop #6 Seattle, WA. https://strbase.nist.gov/pub_pres/AAFS2006_STRanalysis.pdf.

Chambers, G.K., MacAvoy, E.S., 2000. Microsatellites: consensus and controversy. Comp. Biochem. Physiol. B 126, 455–476.

Clark, J.R., Scott, S.D., Jack, A.L., Lee, H., Mason, J., Carter, G.I., Pearce, L., Jackson, T., Clouston, H., Sproul, A., Keen, L., Molloy, K., Folarin, N., Whitby, L., Snowden, J.A., Reilly, J.T., Barnett, D., 2014. Monitoring of chimerism following allogeneic haematopoietic stem cell transplantation (HSCT): technical recommendations for the use of short tandem repeat (STR) based techniques, on behalf of the United Kingdom National External Quality Assessment Service for Leucocyte Immunophenotyping Chimerism Working Group. Br. J. Haematol. 168, 26–37.

Dib, C., Fauré, C., Fizames, C., Samson, D., Drouot, N., Vignal, A., Millasseau, P., Marc, S., Hazan, J., Seboun, E., Lathrop, M., Gyapay, G., Morissette, J., Weissenbach, J., 1996. A comprehensive genetic map of the human genome based on 5,264 microsatellites. Nature 380 (6570), 152–154.

Jeffreys, A.J., Wilson, V., Thein, S.L., 7 March 1985a. Hypervariable 'minisatellite' regions in human DNA. Nature 314, 67–73.

Jeffreys, A.J., Wilson, V., Thein, S.L., 4 July 1985b. Individual-specific 'fingerprints' of human DNA. Nature 316, 76–79.

Li, H., Bai, R., Zhao, Z., Tao, L., Ma, M., Ji, Z., Jian, M., Ding, Z., Dai, X., Bao, F., Liu, A., 2018. Application of droplet digital PCR to detect the pathogens of infectious diseases. Biosci. Rep. 38, 1–8. https://doi.org/10.1042/BSR20181170.

Mika, T., Baraniskin, A., Ladigan, S., Wulf, G., Dierks, S., Haase, D., Schork, K., Turewicz, M., Eisenacher, M., Schmiegel, W., Schroers, R., Klein-Scory, S., 2019. Digital droplet PCR-based chimerism analysis for monitoring of hematopoietic engraftment after allogeneic stem cell transplantation. Int. J. Lit. Humanit. 41, 615–621.

Rakoczy-Trojanowska, M., Bolibok, H., 2004. Characteristics and a comparison of three classes of microsatellite-based markers and their application in plants. Cell. Mol. Biol. Lett. 9, 221–238.

Thiede, C., Bornhäuser, M., Oelschlägel, U., Brendel, C., Leo, R., Daxberger, H., Mohr, B., Florek, M., Kroshinsky, F., Geissler, G., Naumann, R., Ritter, M., Prange-Krex, G., Lion, T., Neubauer, A., Ehninger, G., 2001. Sequential monitoring of chimerism and detection of minimal residue disease after allogeneic blood stem cell transplantation (BSCT) using multiplex PCR amplification of short tandem repeat-markers. Leukemia 15, 293–302.

van der Gaag, K.J., de Knijff, P., 2015. Forensic nomenclature for short tandem repeats updated for sequencing. Forensic Sci. Int. Genet. Suppl. Ser. 5, e542–e544. https://doi.org/10.1016/j.fsigss.2015.09.214.

Further reading

Alvarez-Cubero, M.J., Saiz, M., Martinez-Gonzalez, L.J., Alvarez, J.C., Eisenberg, A.J., Budowle, B., Lorente, J.A., 2012. Genetic identification of missing persons: DNA analysis of human remains and compromised samples. Pathobiology 79, 228–238.

Bader, P., 2019. Documentation of engraftment and chimerism after HSCT. In: Carreras, E., Dufour, C., Mohty, M., Kröger, N. (Eds.), The EMBT Handbook, pp. 143–147.

Fan, H., Chu, J.-Y., 2007. A brief review of short tandem repeat mutation. Dev. Reprod. Biol. 5 (1), 7–14.

Kristt, D., Stein, J., Yaniv, I., Klein, T., 2007. Assessing quantitative chimerism longitudinally: technical considerations, clinical applications and routine feasibility. Bone Marrow Transplant. 39, 255–268.

National Institute of Standards and Technology (NIST), US Department of Commerce, Short tandem repeat DNA internet database. (https://strbase.nist.gov/).

Roper, S.M., Tatum, O.L., 2008. Forensic aspects of DNA-based human identity testing. J. Forensic Nurs. 4, 150–156.

Roewer, L., 2013. DNA fingerprinting in forensics: past, present, future. Invest. Genet. 4, 22.

Index

Note: 'Page numbers followed by "*f*" indicate figures and "*t*" indicate tables.'

K

L

M

N

O

P

R

S

FSC
www.fsc.org
MIX
Papier | Fördert
gute Waldnutzung
FSC® C083411

Kolibri 360 GmbH
Ferdinand-Jühlke-Straße 7
99095 Erfurt, Deutschland
produktsicherheit@kolibri360.de